VETIVER GRASS

A thin green line against erosion

Board on Science and Technology
for International Development

National Research Council

National Academy Press
Washington, D.C. 1993

NOTICE: The project that is the subject of this report was approved by the Governing Board of the National Research Council, whose members are drawn from the councils of the National Academy of Sciences, the National Academy of Engineering, and the Institute of Medicine. The members of the committee responsible for the report were chosen for their special competence and with regard for appropriate balance.

This report has been reviewed by a group other than the authors according to procedures approved by a Report Review Committee consisting of members of the National Academy of Sciences, the National Academy of Engineering, and the Institute of Medicine.

The National Academy of Sciences is a private, nonprofit, self-perpetuating society of distinguished scholars engaged in scientific and engineering research, dedicated to the furtherance of science and technology and to their use for the general welfare. Upon the authority of the charter granted to it by the Congress in 1863, the Academy has a mandate that requires it to advise the federal government on scientific and technical matters. Dr. Frank Press is president of the National Academy of Sciences.

The National Academy of Engineering was established in 1964, under the charter of the National Academy of Sciences, as a parallel organization of outstanding engineers. It is autonomous in its administration and in the selection of its members, sharing with the National Academy of Sciences the responsibility for advising the federal government. The National Academy of Engineering also sponsors engineering programs aimed at meeting national needs, encourages education and research, and recognizes the superior achievements of engineers. Dr. Robert M. White is president of the National Academy of Engineering.

The Institute of Medicine was established in 1970 by the National Academy of Sciences to secure the services of eminent members of appropriate professions in the examination of policy matters pertaining to the health of the public. The Institute acts under the responsibility given to the National Academy of Sciences by its congressional charter to be an adviser to the federal government and, upon its own initiative, to identify issues of medical care, research, and education. Dr. Stuart Bondurant is acting president of the Institute of Medicine.

The National Research Council was organized by the National Academy of Sciences in 1916 to associate the broad community of science and technology with the Academy's purposes of furthering knowledge and advising the federal government. Functioning in accordance with general policies determined by the Academy, the Council has become the principal operating agency of both the National Academy of Sciences and the National Academy of Engineering in providing services to the government, the public, and the scientific and engineering communities. The Council is administered jointly by both Academies and the Institute of Medicine. Dr. Frank Press and Dr. Robert M. White are chairman and vice chairman, respectively, of the National Research Council.

The Board on Science and Technology for International Development (BOSTID) of the Office of International Affairs addresses a range of issues arising from the ways in which science and technology in developing countries can stimulate and complement the complex processes of social and economic development. It oversees a broad program of bilateral workshops with scientific organizations in developing countries and conducts special studies. BOSTID's Advisory Committee on Technology Innovation publishes topical reviews of technical processes and biological resources of potential importance to developing countries.

This report has been prepared by an ad hoc advisory panel of the Board on Science and Technology for International Development, Office of International Affairs, National Research Council. Support for this project was provided by the Division of Environment of the International Bank for Reconstruction and Development (The World Bank), the Soil Conservation Service of the U.S. Department of Agriculture, and the Office of the Science Advisor of the Agency for International Development Grant No. DPE-5545-A-OO-8068-00.

Library of Congress Catalog Card Number: 92-050175
ISBN 0-309-04269-0

PANEL ON VETIVER

NORMAN E. BORLAUG, Centro Internacional de Mejoramiento de Maíz y Trigo Mexico City, Mexico, *Chairman*
RATTAN LAL, The Ohio State University, Columbus, Ohio, USA
DAVID PIMENTEL, Cornell University, Ithaca, New York, USA
HUGH POPENOE, University of Florida, Gainesville, Florida, USA

* * *

NOEL D. VIETMEYER, *Vetiver Study Director and Scientific Editor*
MARK R. DAFFORN, *Staff Associate*

STAFF

F.R. RUSKIN, *BOSTID Editor*
ELIZABETH MOUZON, *Senior Secretary*
BRENT SIMPSON, *MUCIA Intern*

MICHAEL MCD. DOW, *Acting Director*, BOSTID

CONTRIBUTORS

I. P. ABROL, Department of Soils, Agronomy, and Agroforestry, Indian Council of Agricultural Research, Krishi Bhawan, New Delhi, India
EDITHE ALEXANDER, Stanthur and Company, Ltd., Castries, St. Lucia
CHARLES H. ANTHOLT, Agriculture Division, Asia Technical Department, The World Bank, Washington, D.C., USA
ERIK ARRHENIUS, The World Bank, Washington, D.C., USA
JACQUES BARRAU, Laboratoire d'Ethnobotanique-Biogéographie, Muséum National d'Histoire Naturelle, Paris, France
CLIFFORD BELLANDE, CARE-Haiti, Port-au-Prince, Haiti
MICHAEL BENGE, Agency for International Development, Washington, D.C., USA
PAT BOYD, U.S. Department of Agriculture, Glenn Dale, Maryland, USA
TOM BREDERO, The World Bank, Antananarivo, Madagascar
ROBERT CHAMBERS, Institute of Development Studies, University of Sussex, Brighton, UK
GEOFFREY P. CHAPMAN, Wye College, University of London, Wye, Near Ashford, Kent, UK

GABRIEL CHARLES, Ministry of Agriculture, Castries, St. Lucia
CAROL COX, Ecology Action, Willits, California, USA
SELWYN DARDIANE, Director of Forests, Trinidad and Tobago
KITTIE DERSTINE, Soil Conservation Service, U.S. Department of Agriculture, Baton Rouge, Louisiana, USA
MEL DUVALL, George Mason University, Fairfax, Virginia, USA
H. A. ELWELL, Institute of Agricultural Engineering, Borrowdale, Harare, Zimbabwe
V. GIBBERD, E. M. I. Soil and Water Conservation Project, Ministry of Agriculture, Embu, Kenya
M. GRASSI, Instituto de Botánica, Fundación Instituto Miguel Lillo, San Miguel de Tucumán, Argentina
JOHN C. GREENFIELD (The World Bank, retired), Kerikeri, New Zealand
RICHARD G. GRIMSHAW, Agriculture Division, Asia Technical Department, The World Bank, Washington, D.C., USA
BRUCE HANSEN, Herbarium, University of South Florida, Tampa, Florida, USA
R. M. JARROLD, MASDAR (UK) Ltd., Finchampstead, Berkshire, UK
EMANUEL L. JOHNSON, U.S. Department of Agriculture, Beltsville, Maryland, USA
MAURICE JONES, Vetiver Grass Stabilization cc, Management Agricultural Services, Howick, South Africa
R. S. JUNOR, Soil Conservation Service, New South Wales, Chatswood, New South Wales, Australia
W. DORAL KEMPER, Agricultural Research Service, U.S. Department of Agriculture, Beltsville Agricultural Research Center, Beltsville, Maryland, USA
T. N. KHOSHOO, Tata Energy Research Institute, New Delhi, India
MATHIEU KUIPERS, CORDECO-MACA, Cochabamba, Bolivia
GERALD R. LEATHER, U.S. Department of Agriculture, Frederick, Maryland, USA
GILBERT LOVELL, Southern Regional Plant Introduction Station, U.S. Department of Agriculture, Griffin, Georgia, USA
WILLIAM S. LUCE, Miami Beach, Florida, USA
WILLIAM B. MAGRATH, Environmental Policy Research Division, The World Bank, Washington, D.C., USA
MICHAEL McGAHUEY, Agency for International Development, Washington, D.C., USA
MIKE MATERNE, Soil Conservation Service, U.S. Department of Agriculture, Baton Rouge, Louisiana, USA
RAMIRO MATOS MENDIETA, Colegio Real, Universidad Nacional Mayor de San Marcos, Lima, Peru

D. E. K. MILLER, New Zealand Department of Scientific and Industrial Research, Palmerston North, New Zealand
STANDLEY MULLINGS, Stanthur and Company, Ltd., Castries, St. Lucia
ALAN NORDMEYER, Forest and Range Experiment Station, Forest Research Institute, Ministry of Forestry, Christchurch, New Zealand
CALEB O. OTHIENO, Tea Research Foundation of Kenya, Kericho, Kenya
KENNETH PIDDINGTON, Office of Environment, The World Bank, Washington, D.C., USA
C. SITAPATHI RAO, Agriculture and Rural Development Area, Administrative Staff, College of India, Bella Vista, Hyderabad, India
JOHN (MARC) SAFLEY, JR., Soil Conservation Service, U.S. Department of Agriculture, Washington, D.C., USA
CINDY SCHEXNAYDER, Soil Conservation Service, U.S. Department of Agriculture, Baton Rouge, Louisiana, USA
W. CURTIS SHARP, National Plant Materials Laboratory, Soil Conservation Service, U.S. Department of Agriculture, Washington, D.C., USA
HOLLY SHIMIZU, U.S. Botanic Gardens, Washington, D.C., USA
JAMES SMYLE, Agriculture Division, Asia Technical Department, The World Bank, Washington, D.C., USA
S. SUBRAMANIAN, Soil and Water Management Research Institute, Naidu Agricultural University, Thanjavur, India
ANTHONY TANTUM, Vetiver Grass Stabilization cc, Management Agricultural Services, Howick, South Africa
PAUL P. S. TENG, International Rice Research Institute, Los Bãnos, Philippines
PAT THURBON, Department of Primary Industries, Brisbane, Queensland, Australia
PAUL TRUONG, Soil Conservation Research, Department of Primary Industries, Indooroopilly, Brisbane, Queensland, Australia
ARNOLD TSCHANZ, U.S. Department of Agriculture, Hyattsville, Maryland, USA
LY TUNG, Farm and Resource Management Institute, Visayas State College of Agriculture, Pasay City, Philippines
GORO UEHARA, University of Hawaii, Honolulu, Hawaii, USA
J. J. P. VAN WYK, Research Institute for Reclamation Ecology, Potchefstroom University for Christian Higher Education, Potchefstroom, South Africa
RAY WEIL, University of Maryland, College Park, Maryland, USA
PETER WELLE, CARE Haiti, Port-au-Prince, Haiti
P. K. YOON, Plant Science Division, Rubber Research Institute of Malaysia, Experiment Station, Kuala Lumpur, Selangor, Malaysia

Preface

For developing nations, soil erosion is among the most chronic environmental and economic burdens. Many of these nations are in the tropics, where in just a few hours torrential downpours can wash away tons of topsoil from each hectare. Many others are in the drier zones, where swirling winds and flash floods (sometimes from rains so distant they are unseen) can be equally devastating.

By these processes, huge amounts of valuable soil are being lost every day. Worse, the soil accumulates in rivers, reservoirs, harbors, estuaries, and other waterways where it is unwelcome, terribly destructive, and impossibly costly to remove. Erosion is thus a double disaster: a vital resource disappears from where it is desperately needed only to be dumped where it is equally unwanted.

Despite much rhetoric and effort, little has been accomplished in overcoming erosion, at least when viewed from a worldwide perspective. One major reason is that there are few if any solutions that are cheap, appealing, long lived, and suitable for easy adoption over the vast expanses of the Third World that need protecting. Now, however, there may be one.

In the eyes of at least some viewers, a little-known tropical grass, vetiver, might at last offer one practical and inexpensive solution for controlling erosion simply, cheaply, and on a huge scale in both the tropical and semiarid regions. Planted in lines along the contours of sloping lands, vetiver quickly forms narrow but very dense hedges. Its stiff foliage then blocks the passage of soil and debris. It also slows any runoff and gives the rainfall a better chance of soaking into the soil instead of rushing off the slope. Although there has not been much experience with this process to date, the deeply rooted, persistent grass has restrained erodible soils in this way for decades in Fiji, India, and some Caribbean nations.

At least in this limited practice, vetiver appears truly remarkable. The grass itself seems not to spread or become a pest. Terraces rise as the soil accumulates behind the hedges, converting erodible slopes into stabilized terraces where farming and forestry can be conducted, safe from the evils of erosion. Farmers and foresters benefit not only by keeping their soil, but by having flatter land and more moisture for their plants. Countries benefit by having cleaner rivers, unspoiled estuaries, and more water and less silt in their reservoirs.

At present, however, no one knows for sure whether these experiences represent a practical possibility for solving the world's worst erosion problems. The purpose of this report is to make a judgment on this point: to assess vetiver's promise and limitations and to identify any research that may be necessary before this grass can be deployed rationally, widely, and without undue environmental risk. In other words, our particular purpose is to evaluate the ecological advantages and potential risks in employing a grass that may eventually benefit watersheds, forests, and farms throughout the world's warmer zones.

This book, it should be understood, is neither a monograph on vetiver nor a field guide for its use. It is, instead, a scientific audit of the safety and effectiveness of the plant as used for erosion control. Basically, the book reviews existing research and experiences with the grass. It has been compiled from literature, from personal contacts, from site visits, and from information mailed in by specialists in an array of disciplines: environment, agronomy, forestry, soil science, engineering, and others. Many contributors had little or no knowledge of vetiver itself; they served as "devil's advocates." We hope that this exhaustive process has produced an independent, unbiased evaluation that will help scores of countries and organizations judge whether or not to use this plant in their own programs.

The report has been produced particularly for nonspecialist readers such as government ministers, research directors, university students, private voluntary organizations, and entrepreneurs. It is intended primarily as an economic development document. We hope it will be of particular interest to agencies engaged in development assistance and food relief; officials and institutions concerned with environment, agriculture, and forestry in tropical countries; and scientific establishments with relevant interests.

This study is a project of the Board on Science and Technology for International Development (BOSTID), a division of the National Research Council, and is prepared under the auspices of BOSTID's program on technology innovation. Established in 1970, this program evaluates unconventional scientific and technological advances with particular promise for solving problems of developing countries (see page 158). The report continues a series describing promising plant resources that heretofore have been neglected or overlooked. Other titles relating to vegetative improvements for tropical soils and environment include:

- *Tropical Legumes: Resources for the Future* (1979)

- *Leucaena: Promising Forage and Tree Crop in Developing Countries* (second edition, 1984)
- *Mangium and Other Fast-Growing Acacias for the Humid Tropics* (1983)
- *Calliandra: A Versatile Small Tree for the Humid Tropics* (1983)
- *Casuarinas: Nitrogen-Fixing Trees for Adverse Sites* (1983)
- *Firewood Crops: Shrub and Tree Species for Energy Production*, Volumes I and II (1980 and 1983, respectively)
- *Sowing Forests from the Air* (1981)
- *Neem: A Tree for Solving Global Problems* (1992).

Funds for carrying out this vetiver study were made available by the following organizations:

- The Office of the Science Advisor of the U.S. Agency for International Development;
- The Environment and the Agriculture and Rural Development departments of the World Bank; and
- The Soil Conservation Service of the U.S. Department of Agriculture.

Additional funds for printing the report were contributed by the International Tropical Timber Organization as well as by the following departments of the World Bank: Agriculture and Rural Development, Environment, External Affairs, and Asia Technical. The following World Bank country divisions also contributed: South Asia, East Asia, Latin America, and Africa.

How to cite this report:
National Research Council. 1993. *Vetiver Grass: A Thin Green Line Against Erosion*. National Academy Press, Washington, D.C.

Contents

Foreword xiii

Introduction 1

BACKGROUND AND CONCLUSIONS

1 Worldwide Experiences 11

2 Case Studies 32

3 Conclusions 49

TECHNICAL ISSUES

4 Questions and Answers 59

5 The Plant 71

6 Next Steps 84

APPENDIXES

A Great Challenges, Great Opportunities 94

B Other Potential "Vetivers" 113

C Selected Readings 128

D Research Contacts 130

E Biographical Sketches 156

The BOSTID Innovation Program 158

Board on Science and Technology for International Development (BOSTID) 161

BOSTID Publications in Print 162

Foreword

Soil erosion is a quiet crisis, an insidious, largely man-made disaster that is unfolding gradually. In many places it is barely recognized; the soil moves away in such small increments from day to day that its loss is hardly noticed. Often the very practices that cause the greatest losses in the long term lead to bumper crops in the short term, thereby creating an illusion of progress.

Yet erosion is inexorably undermining the economic security of most countries. The changes it brings are chronic and irreversible: lost land; reduced productivity on farms and in forests; floods; silted harbors, reservoirs, canals, and irrigation works; washed-out roads and bridges; and destroyed wetlands and coral reefs, where myriad valuable organisms would normally breed and prosper.

And erosion is literally costing the earth. The soil it carries off now totals 20 billion tons a year worldwide. That represents the equivalent loss of between 5 million and 7 million hectares of arable land. Some of this loss is alleviated by converting forests into farms, so that erosion indirectly also leads to deforestation.

The problems are worst in the warmer parts of the world. There, swelling populations, poor land management, vulnerable soils, and hostile climates add up to a lethal combination that fosters erosion, bringing with it environmental degradation, falling crop yields, rising deforestation, erratic water supplies, and an ever-expanding prospect of dry and dusty rangelands too lacking in soil for crops or even livestock.

Soil erosion is getting worse. In sub-Saharan Africa, for example, it has increased 20-fold in the last three decades as more and more people are forced to move out of the good bottomlands and onto the fragile hillsides. More than one-third of Africa, for example, is threatened with desertification. The world's forests are disappearing 30 times faster than they are being planted. Hillsides stripped of their protective covering of vegetation are rapidly eroding, depositing huge amounts of silt into downstream reservoirs and river valleys. Floods are becoming more frequent—and more severe.

A few facts demonstrate the crisis:

- Morocco has to install the equivalent of a new, 150-million cubic meter reservoir every year just to keep pace with the sediment that is filling up its existing dams.
- Zimbabwe, it is estimated, would have to spread US$1.5 billion worth of fertilizers merely to compensate for the natural nutrients now being swept away by wind and rain every year.
- China loses more than 2 billion tons of soil a year, just from the Loess Plateau. Most is deposited in the Yellow River. And it takes 3.5 billion cubic meters of water to flush every 100 million tons of soil to the sea—water that could be used for productive purposes.
- U.S. farmers must add 20 kilograms of nitrogen fertilizer for every centimeter of soil lost per hectare, just to maintain productivity. Indeed, each year the United States loses $18 billion in fertilizer nutrients to soil erosion.

To avert the global environmental disaster being brought on by soil erosion, it is imperative to take action quickly and on a vast scale. Unfortunately, previous efforts to tackle the problem worldwide—especially in the Third World—were rarely successful over any extensive area. For one thing, some of the conventional techniques employed today are enormously expensive. For another, they rarely generate widespread farmer support—indeed, the farmers often object so vehemently that they have to be threatened with fines or prison to assure compliance. And, wherever the farmers are unmotivated, even the most effective systems soon decline and fall into decay and disuse.

It was with this grim scenario in mind that the staff of the NRC became intrigued by the ideas of two World Bank agriculturists, John Greenfield and Richard Grimshaw. These two had an entrancing vision: a little-known tropical grass called vetiver, they proposed, could provide the answer to soil erosion in the world's warmer regions—and it could do so in a way that would appeal to millions of farmers, landowners, politicians, and administrators. In their eyes, local people would at last be motivated to protect their land and therefore create the solution rather than the problem.

Greenfield and Grimshaw's concept, as well as the reasons behind it, are described in the next chapter. Subsequent chapters highlight the findings of the NRC panel, whose task was to assess the underlying truth of the vetiver idea and to project its promise into the future.

Noel D. Vietmeyer
Study Director

Introduction

In 1956 John Greenfield was handed a problem: to grow sugarcane on the hills of Fiji. With all the flat land fully used, his company had resolved to expand production up the slopes.

The trouble was that the hillsides on that Pacific island nation were too dry for sugarcane—one of the thirstiest of crops—and the soils were too erodible. Putting cane fields on the slopes could be disastrous to the company, the cultivators, and the country.

Everyone but Greenfield's boss knew that there was no chance of farming those lands without bringing on devastating losses of soil. Even Greenfield thought so, but he is a persistent, no-nonsense sort, and he set about testing all kinds of methods in the hope that one of them just possibly might work.

One of the methods involved bulldozing broad dirt barriers (bunds or berms) along the contours. That was the standard process for controlling erosion in the commercial croplands of most parts of the world. Another method involved a coarse grass called vetiver (*Vetiveria zizanioides*), the use of which was virtually unknown.

Greenfield located the vetiver plants growing beside a nearby highway. He had his work crews dig them up, break off slips, and jab the slips side by side into the soil to form lines across the hillsides. It seemed pretty hopeless to expect lines of grass just one plant wide to stop the movement of soil, but he had heard that something similar had been successful in the Caribbean before World War II.

The vetiver slips quickly took root and grew together to form continuous bands along the contour. To everyone's surprise the runoff that normally poured off during tropical storms was slowed down, spread out, and even held back behind the "botanical dams." It took hours to seep through one of the dense walls of stiff stalks (eventually about a meter wide), only to be checked again by the next one down the slope. In consequence, most of the moisture had no chance to rush into the streams below; instead, it soaked into the slopes where it had fallen. More importantly, though, the soil was prevented from leaving the site; it settled out of the stalled runoff and lodged behind the grass walls.

The full significance of this eluded him until the area was hit by one of the highest intensity storms in Fiji history: 500 mm of rain fell in just 3 hours. Water poured over the top of the vetiver hedges "like a tidal wave," but the plants remained in place, undamaged, and unbreached. Of greater consequence, however, was the fact that the water did not start cutting gullies because the bands of vegetation both robbed it of its force and kept it spread evenly across the slopes as it had fallen.

Greenfield's bulldozed earthworks, on the other hand, did not fare so well. Water built up behind them until the weakest spot could hold no longer. Then, the whole dammed-up body of runoff cascaded through the breach, gashing jagged gullies into the erodible slopes below.

After that experience there was little question about which system to pursue. Indeed, the vetiver hedges succeeded so well at both stopping erosion and increasing soil moisture that sugarcane production quickly expanded out of the flats and up the slopes. Some was produced on grades as steep as 100 percent (45°) and without obvious soil loss. Greenfield's boss was right—it could be done!

Greenfield eventually went on to other projects in various countries and spent a long career advising agricultural programs throughout the tropics. In 1985, the World Bank sent him to India as part of an effort to improve productivity in the rundown, rainfed farming areas of four states. In this region, making up much of central India, thousands of small subsistence farms were being ravaged by erosion. Their soils were scarred, thinned, and so quick to dry out that if the rains were interrupted—even briefly—the crops quickly died. When he saw the situation, Greenfield thought of vetiver; if it could benefit a remote Pacific island, perhaps it could benefit India as well.

He described the vetiver concept and related his former experiences to Richard Grimshaw, his new boss in India. It was an opportune time: the project was not working well; the subsistence farms were still eroding and their yields were still low and woefully unreliable. Grimshaw had Greenfield return to Fiji to make a video showing what had happened to his hillside hedges.

To Greenfield's satisfaction, his old grass strips on the northern shore of Fiji's main island were still standing, still protecting the slopes, and still producing excellent sugarcane crops despite three decades of

---

Opposite: Rakiraki, Fiji. What had been smooth, steep (50-percent) slopes turned into tall terraces fronted by dense stands of grass when vetiver hedges were established here in the 1950s. In the years since, the terraces have grown almost 2 m high in some places. Soil losses have been blocked and stable natural terraces formed, all at little or no cost to the farmer. (D.E.K. Miller)

almost complete neglect. Moreover, no one was concerned about erosion—the fact that it had even existed there had been forgotten.

However, the slopes no longer looked as they had when he started. They had been transformed into terraces. Tons of silt had built up behind each line of grass. The area now was composed of strips of flat land behind grassy banks up to 2 m tall.

Back in India, Greenfield's videotape was a big hit, and Grimshaw agreed to give vetiver a try. The plant is native to India (where it is normally called "khus" or "khus khus"), and samples were readily obtained from Karnataka State. A huge nursery—covering 8 hectares—was established near Bhopal, and Greenfield and his colleagues went out to persuade farmers to plant hedges across their sandy, red, eroding land. (The videotape helped immensely, especially because the farmers on Fiji had vividly narrated their experiences in Hindi.)

After 4 years—3 of them involving some of the severest droughts on record—the results were so promising that vetiver seemed like a major breakthrough.

In Maharashtra State, lines of the grass trapped 25 cm of silt in 2 months. On the notorious black soils of Karnataka, silt started forming behind a 3-month-old hedge, and the soil loss dropped from 11 tons per hectare to 3 tons per hectare. On the experimental farm at Sehore University, the average silt buildup behind vetiver lines during the rainy season was 10 cm. Some hedges accumulated as much as 30 cm of soil in a year.

What is more, vetiver held back more than just soil. The moisture that normally gushed off the land in flash floods was also trapped—or at least slowed down. There was no way to measure the exact amount; however, plenty of anecdotal information attested to vetiver's remarkable ability to hold runoff on the slopes. On one droughty Indian catchment, for instance, farmers found they could grow mango trees for the first time. In several villages, the water level in the wells rose dramatically after vetiver hedges were put in. In Gundalpet, for example, farmers had water at 8 m depth; wells in nearby villages were dry down to 20 m. On lands in Andhra Pradesh that had been abandoned 12 years earlier as being too dry for farming, farmers put in vetiver and successfully grew millet and castor once again. In several places, farmers with vetiver lines got a harvest where their neighbors got only crop failures. None of this, it should be noted, was due to vetiver alone. It was the combination of the grass and the contour farming, which it induced and stabilized, that jointly made the difference.

One of the most unexpected successes was on black cotton soils. These "vertisols" cover 10 million hectares of central India and could be a major source of food, but their physical properties are such that they can be used for only a fraction of the year. During the rainy

Gundalpet District, Karnataka, India. On this farm, vetiver hedges have been growing for an estimated 200 years. The hedges are kept in neat, compact lines and are regularly cut to provide fodder, thatch, and other useful materials. (R. Grimshaw)

season, when farming should be most productive, they are too sticky even to walk on. Barriers such as berms or bunds tend to worsen the waterlogging. Vetiver, however, allowed excess water to pass through and didn't upset the drainage. Greenfield noted that, after rain, farmers could walk right up to a vetiver hedge, but couldn't get closer to a berm than 40 m. He was so impressed that he predicted vetiver grass would be *the* answer for bringing India's vertisols into year-round production, resulting in immense benefit to the nation.

Although initially the project staff almost had to force their planting materials onto the farmers, soon the farmers were clamoring for it. Several states—including Orissa, Tamil Nadu, Rajasthan, and Gujarat—also initiated their own vetiver projects. The main constraint to the wider adoption of the method quickly became a lack of nurseries to keep up with the demand for planting materials.

Ironically, after about 4 years of effort, agriculturists in Karnataka stumbled over the fact that many farmers in a particular part of their own state had been using vetiver hedges all along. In fact, throughout the four southern states, people had employed strips of this grass for

St. Vincent's Vetiver

The following is adapted from John Greenfield's report of a trip to inspect the use of vetiver on the Caribbean nation of St. Vincent in 1989. He arrived at a time when a large-scale erosion-control scheme using bulldozers and expensive terracing, to be financed with foreign capital, was being proposed to the people there.

Vetiver grass was introduced to St. Vincent as the major soil conservation measure more than 50 years ago. It was used primarily in the sugar industry to stabilize fields of sugarcane, but found its way all over the island as a stabilizer of road cuttings, driveways, pathways, and tracks along hillsides. Whoever introduced the system did an excellent job, as virtually all the small farmers put their vetiver-grass plantings on the contour. This, together with the resulting contour farming, has saved St. Vincent from the ravages of soil erosion.

Throughout St. Vincent, I stopped to talk to farmers, and the general consensus was, "surely you haven't come all this way to tell us about khus-khus (as it is called locally); we have known how good it is for over 50 years!"

I drove up the leeward coast of St. Vincent with the government official responsible for soil conservation, Mr. Conrad Simon of the Ministry of Trade, Industry, and Agriculture. We visited one of the old sugar plantations (now abandoned) where vetiver-grass

at least 100, and maybe even 200, years. However, they used them primarily to denote the edges of farms. Investigation showed that where the boundaries crossed slopes the hedges held back soil on the upper side. Recognizing the significance of this, a few of the more astute farmers had intentionally planted vetiver across the middle of their slopes. But this was the exception; by and large, the grass's environmental benefits lay unappreciated.

As all these results became apparent, Greenfield, Grimshaw, and most of the other observers became extremely excited. If this little-known grass could benefit some of the worst areas of central India, surely it could benefit other parts of the world. Indeed, here was something that might underpin millions of farms and forests, not to mention watersheds, rivers, reservoirs, harbors, highways, bridges, canals, and other facilities affected by erosion. Perhaps the whole face of the tropics might be changed for the better by the narrow strips of this grass. Perhaps at last there was a way to keep soil on the hills where it was needed and out of the waterways where it was not. More importantly, perhaps it offered a sustainable method of moisture

barriers had formed stable terraces 4 m high. After 50 years, the grass was still active, and there was no sign of erosion. In other areas, vetiver barriers planted on more than 100 percent slopes were providing full protection against erosion and had been doing so for years. The only area where I noticed erosion starting was where people had pulled out the vetiver on the "riser" of a terrace and planted food crops. Major rills had developed down the face, depositing deltas of silt on the terrace below.

When I pointed this out to Mr. Simon, and later discussed it with his colleagues and supervisor, Mr. Lennox Diasley in Kingstown, they all agreed that vetiver grass had given them perfect protection for the past 50 years. So why should they replace it? The main reason seems to be that nobody has ever told them that their system of soil conservation is possibly the best in the world today.

The vetiver system of soil conservation has served St. Vincent well and is a silent partner in the island's farming production. The people have lived with the grass all their lives; indeed, it has become so commonplace that they do not see it. In other words, they have passed those vegetative barriers every day without appreciating the natural terraces that have formed from soil that would have been lost had it been allowed to wash to the sea. But this "silent sentinel" doing its job 24 hours a day, 365 days of the year, has been protecting St. Vincent for the past 50 years. On top of that, it has cost the country nothing.

conservation, a means of "drought-proofing" previously vulnerable rainfed areas.

In the late 1980s, Richard Grimshaw set out to see if vetiver might block the loss of soil and moisture in other parts of Asia.

In China in 1988, for example, he managed to get it incorporated into a very large World Bank program (the Red Soils Project) in Jiangxi and Fujian provinces. The Chinese agriculturists were so captivated with the vetiver hedges' promise that they immediately bought up every vetiver plant they could find and planted lines across 300 hectares of steep slopes. Initially, they used the grass mainly to protect the faces of existing terraces, but they also incorporated small trials to protect tea plantings on steep, smooth, unterraced slopes.

Even on these acidic, infertile lands south of the Yangtze, vetiver grew well. As in Fiji and India, it was able to dramatically slow up soil loss and reduce runoff. This especially impressed the Chinese, who have created vast areas of terracing whose maintenance is very expensive (and getting more so every year). Vetiver hedges promised to slash these costs. Soon, China had set up a vetiver-information network, a collection of vetiver nurseries, and a series of trials in nine provinces.

China was not the only Far Eastern nation to get involved. In the Philippines in 1988, Grimshaw noticed an eroding catchment on the island of Cebu. Nosing around on a hillside nearby, he spied some vetiver and showed the farmer how to dig it up, break off slips, and plant them in lines across the slope. In central Luzon he persuaded the irrigation authorities to test it on farms in the Pantabangan reservoir area. Within 3 months the catchment was starting to stabilize. Since then, it has withstood heavy rainfall (the area receives at least 2,000 mm in a 4-month rainy season) and has turned into terraces that no longer melt away with every rain. Agriculture had become sustainable there for the first time.

In 1990, thanks largely to Grimshaw's enthusiasm, vetiver programs were initiated in Malaysia, Thailand, Laos, Indonesia, Sri Lanka, and Nepal. And soon information started filtering in from places beyond Asia.

Vetiver, it turned out, has been and still is widely used in the Caribbean to keep roadsides and farm fields from washing out. Normally, it is not planted in any organized operation; it is just a commonplace method of erosion protection going back at least 50 years and known to almost everyone. In 1988 Greenfield visited the Caribbean nation of St. Vincent, only to find that the people were about to cut out all the vetiver hedges. There was no erosion problem on the island, they said, so there was no need for these strips of rough grass across their slopes!

Beyond all these surprising experiences, the plant itself turned out to be remarkable. The more Greenfield, Grimshaw, and their colleagues learned about it, the more amazed they were. Some of the features that came to light were indeed hard to believe.

One clump of vetiver remained submerged beneath muddy floodwaters for 45 days and still emerged alive and apparently unscathed. Others withstood fire. In Fiji, for instance, the practice of burning cane trash after the harvest each year does no lasting damage to the rows of vetiver weaving through the fields. In fact, creeping ground fires usually stop dead when they meet a dense wall of the grass in its green state.

Vetiver is immune to many other Third World hazards as well. The crown of the plant occurs slightly below the soil surface so that grazing goats or even trampling cattle do no lasting damage. Moreover, the mature foliage is so tough and coarse that even with all the cattle roaming the Indian countryside, the plant is never destroyed. This is important for an erosion control crop that, if it is to work effectively, must stay in place for years, even in the presence of hordes of hungry animals. On the other hand, if the vetiver plants were cut back during the growing season, the masses of new green shoots that quickly arose could be fed to livestock.

Vetiver also grew roots amazingly fast. In a special glass-fronted box that Greenfield built in New Delhi, the roots reached as deep as 1 m just 2 months after planting, even though it was winter. In 3 months they were more than 2 m deep.

Of added significance was the fact that the roots grew almost straight down. There were few lateral surface roots. This seemed to explain the widespread observation that sugarcane and other crops would grow right up to a vetiver hedge, seemingly without interference and loss of yield.

Above ground the plants also grew rapidly. In certain sites they reached 2 m tall in just a few weeks. On the other hand, they spread only slightly. Indeed, the bases expand so little that in certain parts of India vetiver hedges are legally accepted as property lines. In Nigeria, too, the surveyor general has in the past permitted vetiver as a legal boundary marker.

All this may seem strange, but there is more. Vetiver grows so densely that, at least according to various informants, it can block the spread of weeds, including some of the world's worst creeping grasses: couch, star, kikuyu, and Bermuda. In Zimbabwe, for example, tobacco farmers reportedly plant vetiver around their fields to keep kikuyu grass from creeping in. In Mauritius, sugarcane growers rely on vetiver to prevent Bermuda grass from penetrating their fields from adjacent roadsides.

Some claim that vetiver also blocks the spread of animals. Many in India, for example, say that snakes will not go through a stand of vetiver because, they believe, the sharp-edged leaves cut the snakes' bodies. This probably mistaken belief leads people to walk confidently where there is a vetiver hedge. In China, a farmer planted vetiver grass around his pond: within six months it had formed an almost impenetrable hedge that not only corralled his ducks but also provided fodder and mulch.

Not the least of these observations was that vetiver can grow in an amazing range of soils. In fact, the plant seemed to thrive so well in adversity that it was hard to find a place where it would not survive. Even in quartz gravel, with little fertility, Greenfield was able to make it grow, after a fashion. And, most impressive to a plant scientist, in Sri Lanka some vetiver grows in bauxite, a material that is toxic to almost every other species of plant. (The farmers there actually used crowbars to plant the vetiver into solid bauxite.)

It can grow in an amazing range of climates as well. Both Fiji and central India lie in the monsoonal tropics, but soon it became clear that vetiver could grow in many other climes. In India, for example, it is found in the rainforests of Kerala, the deserts of Rajasthan, and the frost zones of the Himalaya foothills. (It is even found near Kathmandu in Nepal.) It occurs on some coasts, growing in the direct path of salt spray. All in all, it seemed that vetiver grass could thrive under very wet (more than 3,000 mm) and very dry (less than 300 mm) conditions, and perhaps everything in between.

Vetiver also proved well adapted to temperature. It thrives in Rajasthan, where temperatures reach as high as +46°C, and it survives in Fujian, China, where winter temperatures have reached -9°C. It is certainly surprising that a tropical grass can withstand such cold, but one enthusiast even planted it on a ski slope in Italy, and there, north of Rome, it has (perhaps miraculously) made it through several winters.

With all these discoveries about a species almost unknown to the world, it is not surprising that Greenfield and Grimshaw were enthusiastic. Eventually, through word of mouth and the World Bank's *Vetiver Newsletter* and its handbook, *Vetiver: a Hedge Against Erosion*, others also were caught up in the fervor. Indeed, the excitement surrounding vetiver grew so much and the implications of using it seemed so astounding that many outsiders queried the sincerity (not to say sanity) of those who were discussing it with such zeal.

It was the skepticism surrounding the extravagant claims about the plant that led to this report. The World Bank, the U.S. Agency for International Development, and the U.S. Soil Conservation Service asked the National Research Council to evaluate the claims, the practical reality, and the implications behind the vision of John Greenfield and Richard Grimshaw.

The rest of this report highlights our findings and conclusions.

1

Worldwide Experiences

The experiences described in the previous chapter seem to promise a new and perhaps invaluable technique for controlling erosion in the tropics. However, it should not be thought that John Greenfield and Richard Grimshaw were the first to promote vetiver. On the contrary, this plant is one of the better known crops of the tropics. It is only for erosion control that their efforts stand out.

For several centuries vetiver has been commercially cultivated for the scented oil that can be distilled from its roots. This is a treasured ingredient in some of the world's best-known perfumes and soaps and, largely because of its potential as an export commodity, vetiver can be found in at least 70 nations (see table next page). Indeed, during the last century, wherever British, French, and other colonial administrators were assigned, they typically established test plots of essential-oil crops—vetiver, citronella, and lemongrass, for instance. In most cases, vetiver still remains in those plots, but only a handful of countries produce the oil commercially (see sidebar, page 78).

Despite this, however, in most places use of the grass to halt soil loss is virtually unknown. Nonetheless, the plant's ability to control erosion is not new. Before World War II, some tropical countries deliberately planted vetiver hedges as contour barriers. This was particularly true in the sugarcane fields of the British Caribbean. However, this technique was essentially forgotten during the disruptions of the war and of the independence that soon followed in many nations. With both the war and the collapse of colonialism, many agricultural advisors left the tropics, taking the knowledge of this technique with them. By the 1960s, only a few people in the former colonies remembered that vetiver could stop erosion.

But vetiver is so persistent that even where it has been abandoned, it has continued to survive for decades or even centuries. For this reason, therefore, the plant can be found throughout the tropics as well as in a few other (sometimes completely unexpected) areas.

Below, we highlight the findings from a correspondence campaign

TABLE 1 Countries Where Vetiver is Currently Known To Exist

Africa	Asia	Caribbean
Algeria	Bangladesh	Antigua
Angola	Burma	Barbados
Burundi	China	Cuba
Comoro	India	Dominican Republic
Central African Republic	Indonesia	Haiti
Ethiopia	Japan	Jamaica
Gabon	Malaysia	Martinique
Ghana	Nepal	Puerto Rico
Kenya	Pakistan	St. Lucia
Madagascar	Philippines	St. Vincent
Malawi	Singapore	Trinidad
Mauritius	Sri Lanka	Virgin Islands
Nigeria	Thailand	**Pacific**
Rwanda	**Americas**	American Samoa
Réunion	Argentina	Cook Islands
Seychelles	Brazil	Fiji
Somalia	Colombia	New Caledonia
South Africa	Costa Rica	New Guinea
Tanzania	French Guiana	Tonga
Tunisia	Guatemala	Western Samoa
Uganda	Guyana	**Others**
Zaire	Honduras	France
Zambia	Paraguay	Italy
Zimbabwe	Suriname	Spain
		USA
		USSR

aimed both at locating the grass and at roughly assessing the worldwide experiences with it.[1]

ASIA

Vetiver is an Asian plant, probably native to a lowland, swampy area north of New Delhi in India. It has therefore been known to Asians longer than to anyone else.

India

In India, vetiver, or *khus* as it is more commonly known, has been used since ancient times and is recorded as a medicinal plant in the Ayurvedic era. However, it has been appreciated mainly for its

[1] This campaign was conducted during 1990 and 1991. In a number of cases vetiver had been little recognized and never recorded in these locations. The plant looks so ordinary that it is, in a sense, an "invisible" cohabitant—listed neither in the compendiums of native plants nor in the lists of commercial resources.

fragrant roots. People weave these roots into mats, baskets, fans, sachets, and ornaments. They also weave them into window coverings that freshen the air of thousands of village homes with a sweet and penetrating scent.[2] Oil from the roots is also used in perfumes. Not only is it pleasantly fragrant, it takes so long to evaporate from the skin that perfumers include it in their soaps and scents to give them "persistence."

The living grass has also long been known as a useful soil binder. However, it was mostly planted around rice paddies, along rivers, and beside canals and ponds to strengthen the banks and keep the land from collapsing into the water.

To line it out across the hill slopes is, for most places, a new and innovative concept. However, since 1987 a number of trials in research stations and farmers' fields have been carried out under the aegis of several of India's state governments, largely with funding from the World Bank. The data is preliminary and is usually not statistically significant, but it is instructive nonetheless.

Some examples follow.[3]

Karnataka The state of Karnataka has taken up vetiver for watershed conservation with considerable enthusiasm. Traditionally, the state's farmers have used the grass as boundary plantings, and, although some have placed hedges across the middle of their land—an indication that they understand its value for erosion control—its use to conserve soil and moisture is essentially new.

Among the results of this exploratory research are some from Kabbanala.[4] Even in their first year, the partially formed hedges held back 30 percent more rainfall runoff than graded banks, 47 percent more than conventional contour cultivation, and 24 percent more than hedgerows of leucaena.[5] In addition, they held back 43 percent more soil than graded banks, 74 percent more than contour cultivation, and 54 percent more than leucaena hedges. Moreover, apparently because of the improved soil-moisture levels, the vetiver hedgerows boosted crop yields 6 percent more than those on graded banks, 26 percent more than those on contour cultivation, and 10 percent more than those behind leucaena hedgerows.

---

[2] Typically, people throw water on these vetiver "shades," which brings out the fragrance and both cools and scents the breeze wafting through.

[3] For an earlier experience with vetiver, near Lucknow in Uttar Pradesh, see next chapter.

[4] This research was conducted by A.M. Krishnappa. The hedges were placed at 1-m vertical intervals on slopes of less than 5 percent.

[5] This fast-growing tree, *Leucaena leucocephala*, is described in a companion report, *Leucaena: Promising Forage and Tree Crop for the Tropics*, National Academy Press, Washington, D.C., 1984.

Maharashtra and Madhya Pradesh On the black cotton soils in these two states, vetiver has also grown well. It holds back soil and some moisture but (unlike bunds) does not create pools of standing water on these heavy, ill-draining clays. This is important because these soils are so difficult to work with that at present they are only partially used: in the wet season they get too waterlogged; in the dry season they crack open. Bringing the black cotton soils into fuller use could boost India's food production because there are millions of hectares of them.

Tamil Nadu In Tamil Nadu there has been no formal governmental support thus far, but the Regional Research Station at Aruppukkotai[6] began testing vetiver in 1987. Another grass and two shrubs were later added to the trial.[7] Both grasses grew into hedges far more quickly than the shrubs, and both also proved better at trapping moisture. Vetiver was the best of all. It retained between 3 and 9 percent more moisture than the other plants. Soil behind it contained 26 percent more moisture than that on the control slope, which lacked protective hedges.

All in all, vetiver is starting to be accepted for erosion control. The state's department of agriculture has taken up large-scale multiplication of vetiver. Extension agents are being traincd. And contour hedges are being established in selected watersheds. Ultimately, the state plans to protect much of its dryland areas with vetiver.

Andhra Pradesh The government of Andhra Pradesh has planted vetiver in several watershed-management projects. The results, however, are not completely satisfactory. Because of the state's semiarid climate, the plants take at least three years to grow together into fully functional hedges. Farmers, therefore, are as yet unconvinced of vetiver's value. Indeed, some, concluding that the plant will never do them any good, have plowed it up. Nonetheless, more than three out of four are waiting to see.

Nepal

With its steep slopes and rushing rivers, Nepal is one of the most erosion-prone nations. Vetiver has traditionally been used in the lowland area known as the Terai, but only to stabilize the banks of waterways. Today, it can be found verging many irrigation canals, especially in locations where people tend to walk. The roots are commonly harvested and made into hairbrushes, among other things.

---

[6] Information from S. Subramanian.

[7] The grass was kolukattia grass (*Cenchrus glaucus*); the shrubs were hedge lucerne (*Desmanthus virgatus*) and leucaena (also known as subabul in India).

Also, basket weavers prefer vetiver stem; they say it holds paint and keeps its color better.

Given the results in neighboring India, much interest in vetiver has recently surfaced in Nepal. Both private organizations and government agencies have established nurseries and are growing the grass. They hope, primarily, to use it to protect the front lip of terraces (of which there are vast numbers), but they see it also as a possible way to stabilize roadsides and to protect (even "renovate") landslide areas. One group is trying to save a small dirt airfield by planting vetiver along the banks of a hungry river that is slowly eating it away.

All efforts so far have been in the Terai and nearby areas. Although it occurs commonly in the vale of Kathmandu, vetiver probably cannot

Gundalpet, India. In this part of Karnataka State, vetiver hedges are a traditional part of farming practice. Even on fairly level land, they can accumulate impressive amounts of soil behind them. (R. Grimshaw)

withstand the frigid winters of the uplands, where perhaps the most devastating erosion is occurring.

Sri Lanka

Vetiver traditionally has been used to stabilize some slopes and terraces in tea plantations around Kandy. However, its true potential for Sri Lanka lay unappreciated until 1989 when Keerthi Rajapakse, a retired Assistant Conservator of Forests, helped establish nurseries to supply vetiver planting material to farmers. Rajapakse was elated to find that tobacco cultivators accepted this vegetative contour method with alacrity: soil washing out of hillside tobacco fields is considered to be one of Sri Lanka's major environmental problems.

The plant's ruggedness is almost legendary in Sri Lanka. It is here that people have used crowbars to plant it into bauxite soils. Also, farmers say that couch grass (a creeping weed almost impossible to keep out of crops) cannot penetrate a vetiver hedge.

Indonesia

Although Java is the leading producer of vetiver oil, almost nowhere in Indonesia has the grass been used in erosion control. Ironically, in some places people are convinced that it actually *causes* erosion. This is because harvesters often rip out the roots (for the oil they contain), leaving behind trenches that foster the severest of soil losses. This has been such a problem that vetiver cultivation has been prohibited in parts of Java. It is, however, a problem of irresponsible harvesting, and is irrelevant to hedges left in place as erosion barriers.

Vetiver is now being established in some of the other islands (in Kalimantan, for example) with some success.

Philippines

Vetiver can be seen throughout the Philippines. It is reported that a few areas have traditionally cultivated it to control erosion—especially around ponds to keep silt from washing in. Nevertheless, the plant is hardly known to Filipinos at large.

Today, however, that is changing. A major project to restabilize the roads destroyed by the 1990 earthquake in Northern Luzon is relying on vetiver. A number of farmers in both Northern Luzon and Central and Eastern Visayas have put in contour vetiver hedges to stop soil losses. The International Rice Research Institute (IRRI) at Los Baños is studying the grass as a way to reduce erosion in upland rice fields and to reinforce the bunds around the paddies in the lowlands. Visayas

State College of Agriculture has found that vetiver grows well even on poor and very acid upland soils where little else can survive.[8]

China

Although its efforts are just beginning, China is among the nations most active in studying vetiver. Massive projects have been started in a dozen soil-conservation areas. In part this is because of the promise of the early results, but it is also because soil and moisture conservation are among China's national priorities. The country annually loses as much as 400 tons of soil per hectare in certain areas. As a result, the Yellow River is supposedly the most silt laden in the world, and the dry-season levels in the Yangtze are dropping year by year as the ever thinner layer of soil on the hills absorbs less and less moisture. Thus, it is little wonder that there is an almost desperate grasping at what might possibly be a low-cost solution that can be installed on a large scale with local labor and resources.

Vetiver was introduced to China in the 1950s as a source of aromatic oil; when the oil prices dropped, the plant was abandoned. The impetus to test it for erosion control began only in 1988, when it was planted for stabilizing terraces of citrus and tea in Fujian and Jiangxi provinces. Farmers noted that the young citrus and tea plants seemed to grow better (perhaps because of wind protection or increased moisture) and the terraces no longer washed out.

In some areas, vetiver products are already widely popular. In at least one location, prunings from the contour hedges are sold to dairy farmers as a feedstuff and are replacing rice straw as bedding for animals. (The farmers like vetiver straw for bedding not only because it is cheap and rot-resistant, but because it frees up valuable rice straw for sale or for plowing back into the paddy.) Elsewhere, the leaves are employed for mulch.

Although China's Ministry of Water Resources has set up vetiver trials and demonstrations on some severely gullied lands, it is currently more interested in protecting terraces. Nowhere is this interest stronger than in the red-soils area, where engineered terraces have a long record of failing during the intense downpours that occur every few years.

Overall, the results in China have been so promising that during the next few years vetiver will be planted extensively in five provinces. Moreover, Chinese researchers have started their own vetiver-information network, and the government has sponsored a number of vetiver conferences to speed up the exchange of information and results.

Already it has been learned that vetiver can be used south of the

---

[8] Information from Ly Tung. The acidity is so great that the pH was below 5.

Yangtze. However, stands have been established as far north as 36°N in Shangdong Province, where winter temperatures drop to -8°C.[9] So far, there have been few "scientific" trials, but in one carefully documented case vetiver hedges decreased the amount of water running off the slopes by half.

AFRICA

Although the plant is not well known in Africa, it actually can be found growing in scattered locations from Cairo to Cape Town. Also, there is at least one native species that is an African counterpart to this Asian plant. A few African countries have already embarked on exploratory trials with one or both of these species.

Several examples follow.

Kenya

Erosion has long been serious in Kenya, especially in the highlands where the rainfall is heavy and the soils erodible. Vetiver occurs in several locations there, although it is mostly used as an ornamental. It was probably introduced to produce vetiver oil, but it is thought that some was also brought in by a German coffee grower secking to protect his eroding land. Recently, a government soil specialist[10] found vetiver established on terraces in coffee country in the Machakos District as well as on a dam wall on a farm near Thika. He reports that the plants "are still effectively protecting the soil after at least a decade of neglect."

Vetiver, in principle, could be extremely valuable in the highlands—in plantations of tea, coffee, and pyrethrum, as well as in family gardens and along the sides of roads and tracks. Trials began in 1990 and, despite the short time, the grass is already showing promise.[11]

Tanzania

Although now all but unknown in Tanzania, tea planters used vetiver as an erosion-control barrier before World War II. Indeed, an old monograph on tea-growing there makes the following comment: "Grasses are usually detrimental in a high degree to the growth of tea, with apparently one exception, as far as trials and experience goes.

[9] Although vetiver overwintered and is doing well in its second season, this northerly area is probably beyond the latitude where the plant can be used with confidence.

[10] V. Gibberd of the Ministry of Agriculture's E.M.I. Soil and Water Conservation Project.

[11] Information from C.O. Othieno. One interesting possibility being explored is whether vetiver can keep mole rats out of fields and plantations. These burrowing rodents damage crops and trees by chewing on the roots. It is thought that vetiver's dense wall of oil-laden roots might deter them.

Pantabangan Watershed Catchment, Northern Luzon, the Philippines. This two-year-old hedge was planted for the purpose of erosion control. Even in the heavy shade under these trees, the vetiver plants are growing well. The hedges not only reduce soil loss, they also help retain runoff moisture. As a result, the trees grow better and in new tree plantings the number of seedlings that survive can rise dramatically. (R. Grimshaw)

This exception is khus-khus grass [vetiver]. With its close stiff blades it is successfully used . . . in contour hedges. Its tussock habit favors its usefulness as its root system appears to be reasonably restricted in range. It needs to be kept under control by cutting."

Vetiver established on Mount Kilimanjaro, near Arusha, in the final decade of the last century still remains.

Mali

The vetiver highlighted in this report (*Vetiveria zizanioides*) is probably found throughout West Africa, but far more common at present is an African counterpart, *V. nigritana*. Whether it, too, will prove useful in erosion control is uncertain, but it has many interesting qualities nonetheless.

For example, *V. nigritana* occurs across the floodplains of the internal Niger Delta, a vast region (about 20,000 km^2) that is inundated for half the year. Here, it occurs mainly on higher ground, and after the flood retreats it becomes lush and green and the herds are then given access to it according to traditional grazing rights. Vetiver (and a mixture of other species) thus provides crucial grazing to the people of the area. Animals take only the young sprouts, but after these are gone, the farmers burn the area to induce a flush of new growth.

For all its importance here, vetiver is not a fine feedstuff. Annual yields vary from 2 to 10 tons of dry matter, depending on the previous flood levels. The International Livestock Centre for Africa (ILCA) found that the whole plant had a digestibility coefficient of 35–40 percent when cut about 2 months after the flood retreated or after being burned—considerably less than the 60-percent level for the other native grass species in the area.

Despite these indifferent figures, however, this African vetiver occurs extensively, and it resists the harsh cycle of floods and fires better than almost any other species there.

Burkina Faso

In neighboring Burkina Faso, *Vetiveria nigritana* is also important. Indeed, certain tribes rely on the plant throughout their lives. For example, the Bozo (who dwell along the vetiver-covered shores of lakes as well as the Niger River) weave their huts out of vetiver grass or palm fronds.

Nigeria

Vetiver is commonly seen growing wild in Nigeria and probably this is mostly the African species, *Vetiveria nigritana*. Until recently, it was not used for erosion control but now that is changing. Scientists in Anambra are taking the lead.

A state in the southeast, Anambra is probably the area hardest hit by erosion. It has extensive "badlands," so-called because of steep slopes and gullies that engulf the land and everything on it. Erosion is so severe that in just the last 10 years some 220 towns have lost property worth nearly 6 billion naira ($755 million) to 530 greedy gullies. More than 150 people have died from the resulting floods, slips, and cave-ins.

Alarmed, the state government in the early 1980s assembled a team of soil scientists and engineers from its ministries and universities and told them to find out how to halt the further spread of this menace. Initially, many concrete embankments were built, but to little avail. Then some of the scientists came up with the idea of planting strips of grass. Twenty-seven different species were tested; vetiver proved the fastest growing and the most effective. (Initially, *V. nigritana* was used, although *V. zizanioides* is now used as well.)

The researchers were excited by these results; however, the local people—even those most afflicted by erosion—remained unconvinced. The plants looked too weak for such a big job. The scientists' frustration rose and rose until, in March 1990, Britain's Prince of Wales came to town. With his intense interest in conservation, Prince Charles gladly agreed to plant some vetiver personally. He thereby launched what is officially called "The British Council/Anambra State Project on Erosion Control," but is more commonly known as "Project Vetiver."

Thanks to the heir to the British throne, vetiver planting took off in earnest. "Prince Charles brought prestige to bear on the whole thing," noted Anthony Chigbo, secretary and project engineer of Project Vetiver. "We knew that the long strong roots would go deep and hold the soil in place, but we couldn't convince the average person. They believed that things are good only when they are costly!"

Within a year, field workers were reporting favorable results. Indeed, the Nigerian scientists and their British helpers now jointly predict that vetiver will significantly benefit the environment of Anambra.

Ethiopia

Indian scientists reputedly introduced vetiver to some Ethiopian coffee plantations in the early 1970s. Today the grass is used particularly in Jimma and Kaffa provinces, where small informal nurseries of it are often seen along the roadsides. The Ministry of Coffee and Tea has promoted its use there for at least 10 years.

Ethiopians mainly use vetiver to protect the edges of contour drains, but the plant is becoming increasingly popular as an ornamental around houses. In addition, local farmers have found that the foliage makes excellent mulch, and they say that (compared to napier grass, for example) it is easier to manage because it does not seed or take root when they spread it on their gardens or fields. Pine needles, which were traditionally laid on the floor during coffee ceremonies, are now commonly replaced by vetiver leaves. Vetiver straw is also gaining popularity as a thatch and as a stuffing for mattresses because it resists rot and lasts longer than other straws.

One advantage, widely believed in Ethiopia, is that Bermuda grass and couch grass cannot invade fields through a vetiver hedge. Indeed, the local Amharic name for vetiver means "stops couch grass."

Zimbabwe

Although vetiver is only now getting its first serious trials as an erosion control in Zimbabwe, Mauritian settlers in the Chiredzi area (hot, subtropical lowveld, 600–900 m altitude) have reportedly used it for years to reinforce the banks of irrigation canals. They also took it to the Chipinga area in the Eastern Highlands (warm, subtropical hill country, 900–1,200 m altitude), where a number of coffee planters use it to protect their terraces. Vetiver has also been planted across wet drainage flats (vleis), where it blocks the runoff, thereby keeping the soils moist for months into the dry season.

It has also been reported that tobacco farmers have found that hedges of vetiver around their fields keep out creeping grass weeds, such as kikuyu and couch.

Other African Nations

South Africa Vetiver is cultivated to a limited extent in South Africa and is used as a hedge plant, particularly in Natal where it is used mainly by Mauritian sugarcane growers and is commonly referred to as "Mauritius grass." A company formed for the purpose of putting in vetiver hedges has recently established trials throughout the country (see next chapter).

Madagascar Erosion is such a tremendous problem in Madagascar that farmers have rapidly come to recognize vetiver's usefulness. The grass is being planted, largely under a World Bank initiative. (The Bank's local representative is widely known as "Monsieur Vetivère.") Farmers have found that vetiver fits well with their traditional technique of torching their fields each year. Whereas most other erosion-control plants are destroyed, the vetiver is all but unaffected. Another reputed advantage is that vetiver does not harbor rats. More details are given in the next chapter.

Botswana In Botswana, vetiver (*V. nigritana*) is known to occur in the Okavango swamp area, but as of 1991 no erosion-control projects have been reported with either *V. nigritana* or the Indian species (*V. zizanioides*).

Malawi Vetiver is said to have been used in Malawi for 50 years to stabilize sugarcane.[12]

Mauritius Sugarcane growers on Mauritius rely on vetiver and take it for granted. "Any sugarcane grower in Mauritius will tell you that vetiver is used both for erosion control as well as some sort of barrier to prevent noxious weeds such as Bermuda grass from penetrating fields from roads," writes Jean-Claude Autrey, head of the Plant Pathology Division of the Mauritius Sugar Industry Research Institute. "The abundant root system of vetiver is ideal for this purpose."

Zaire The current situation is unreported, but in the 1950s vetiver was commonly found as an ornamental plant and as a border against erosion. Vetiver hedges were sometimes used to fix terraces in place on plantations of cinchona, for example.[13]

Central African Republic A solitary note in the Kew Herbarium Collection reports that vetiver is used for stuffing mattresses in the Central African Republic.

Rwanda For more than 30 years vetiver has been used in Rwanda's coffee plantations, apparently for protecting terraces.[14]

Gabon According to one report, the grass was planted along ditches and roadsides to conserve the soil and delimit field boundaries.[15]

Ghana Vetiver is a common hedge plant in Ghana. It is used particularly along the edges of roads, gardens, and cultivated fields. It is said to prevent a weed called "dub grass" (*Desmostachya bipinnata*) from invading. The leaves, in their young state, are used as cattle fodder.

Tunisia Apparently, Europeans introduced the grass into Tunisia, and it has been planted alongside pathways to conserve the soil.[16]

---

[12] Information from M. Jones.

[13] M. Van den Abeele and R. Vandenput. 1956. *Les Principales Cultures du Congo Belge*. Publication de la Direction de l'Agriculture, de Forêts, et de l'Elevage, Brussels.

[14] Information from V. Nyamulinda.

[15] A. Raponda-Walker and R. Sillans. 1961. *Les Plantes Utiles du Gabon*. Editions Paul Lechevalier, Paris.

[16] J. Trochain. 1940. Contribution a l'étude de la végétation du Senegal. *Memoires de l'Institut Français d'Afrique Noir*.

CARIBBEAN

The Caribbean is one of the regions where vetiver is best known. Originally, "khus-khus grass" (as it is mostly called) was brought in from India. Today, it is commonly cultivated to avoid soil wash and the invasion of weeds. It is also an ornamental. The dried roots are sometimes used as an insect repellent, to protect clothes from moths, for example.

St. Lucia

Vetiver has been used for erosion control on St. Lucia for at least 40 years (see next chapter). It is still widely valued, particularly on the wetter (southwest) side of the island. The foliage is harvested for mats and handicrafts. In earlier times, thatch was the main vetiver product. These days, however, the only places still employing it are said to be tourist facilities for visiting Americans.

Vetiver is often planted around buildings during construction. For example, the Hess Oil Company recently protected slopes around the schools it built near its refinery by using lines of vetiver.

Trinidad

Seemingly, it was the University of the West Indies[17] that originally recognized vetiver's usefulness for soil conservation. As a result, the plant can now be seen all over Trinidad. Mainly it is planted alongside roads. Indeed, it is vetiver that stabilizes the embankments of many of Trinidad's roads.

At the university research station at St. Michaels, for example, vetiver's potential for stabilizing roadsides under the worst possible conditions can be seen. The road has been bulldozed into the side of the hill, and the resulting debris—subsoil, rock, and shale—is so bad it can hardly be called soil, yet the plants are growing actively and there is no sign of erosion.

Higher up the hills on this research station are found vetiver barriers established in old, eroding, "slash-and-burn" areas on slopes well over 100 percent (45°). Here vetiver must compete with jungle regrowth, heavy grass, and vines, but it has held its own and is doing an excellent job of holding these hillsides together.

These days, Trinidad's forest service is showing renewed interest in vetiver.[18] In Maracas Valley, for example, it has planted fruit trees

---

[17] Then known as the Imperial College of Tropical Agriculture, it has long been a globally renowned research center for the study of tropical crops. Apparently, the word of vetiver's utility in erosion control spread in this way.

[18] Information from Selwyn Dardiane, Director of Forests. This activity predates the rise in recent interest stimulated by the World Bank.

behind vetiver contour barriers. On extremely steep slopes, as well as on terraces, one can see deep deposits of organic matter trapped behind the grass hedges. Mango trees are obviously benefiting (probably from the extra fertility as well as the deeper percolation of moisture); those away from the barriers are poor by comparison.

Haiti

Because Haiti is the world's second biggest supplier of vetiver oil, the plant is very well known there. Unfortunately, however, in this country of impoverished soils, it has not been widely employed for erosion control. Its extreme robustness can easily be seen because it commonly occurs on the worst possible sites. Even when people are trying to produce vetiver roots commercially, they usually employ the least valuable ground, much of it so worn out that nothing else can survive.

However, only southern Haiti grows vetiver for oil. In the northern part of the country, people leave it in place and use it for soil conservation. "Vetiver works," notes one admiring Haitian. "The minute you remove it the banks fall down."

A good example of vetiver's abilities can be seen on the road from Port-au-Prince to Cap Haitien. Where it cuts through the hills, the embankments on both sides are lined with vetiver. The terrace effects have successfully stabilized these banks, an amazing feat considering their steepness and erodibility.[19]

Haitians like several of vetiver's features: it can withstand animals (which eat leucaena, for example), is easy to propagate, is drought hardy, and stays in place with minimum maintenance. Moreover, they point out that people can walk over vetiver without damaging it; doing that to leucaena generates a gap.

One widespread opinion that vetiver impoverishes soils has been debunked. Barrenness does occur in some vetiver areas, but hardly because of the plant. The hillsides had first been damaged to the point where farming was no longer possible, then turned over to goats, and only after the goats could find nothing else to eat did the farmers put in vetiver. For such sites vetiver was merely the last straw.

So far, not much vetiver has been used to stop farm erosion.

Barbados

Vetiver is so common on Barbados that it is part of the landscape, and is much appreciated. A recent tourist brochure, for example, gushes over the plant:

---

[19] Information from Clifford Bellande and Peter Welle.

Huge clumps of this bushy grass form continuous borders along most of the country roads in Barbados, providing a soil-erosion barrier around fields of its cousin, the sugarcane. This, however, is perhaps a lesser virtue to most people, the most gracious being the gorgeous smell of this rare fragrance of the tropics. The oil extracted from the roots is used to make a truly Caribbean scent—Khus Khus by Benjamins of Jamaica—available at all fine perfume counters.

The dried roots are also used to make unique souvenirs. These include: clothes hangers covered in khus-khus roots and bound in ribbons. (They give your cupboard a glorious, long-lasting scent, are thick and soft for hanging clothes on, and look very special.)

The leaves of the grass were used extensively to make thatch roofs in early days, but now only as a decorative, "typically tropical" feature. Examples can be seen at Southern Palms Hotel and Ginger Bay. In crafts, khus-khus grass is woven to make Dominican-style rugs at the IDC Handicraft Division. These can be bought by the square for wall-to-wall covering. Ireka, a Rastafarian girl from Mount Hillaby, uses khus-khus grass with balsam to make a whole range of distinct baskets which she sells at her shop at Pelican Village. Roslyn of Barbados uses the grass to make wall hangings and lampshades which are sold at Fine Crafts in the Chattel House Village, and at Articrafts in Norman Centre, Broad Street.

SOUTH AMERICA

Vetiver can probably be found throughout tropical South America, but this is the part of the tropics where it is probably least known. Few South Americans are aware of its presence.

Argentina

The plant is cultivated in the provinces of Chaco and Misiones and it is grown even as far south as Buenos Aires. Its roots are extracted and the essential oil used in perfumes. In Misiones it is employed in thatching ranchos.[20]

---

[20] Information from M. Grassi. The plant's common names in Argentina are both vetiver and capia.

Bolivia

Although there has been no research on erosion hedges, the plant is known in Bolivia. Indeed, a project to explore this use has recently been initiated jointly by the Ministry of Agriculture and Campesinos and the Corporation for the Development of Cochabamba. Vetiver plants are now growing well in a nursery in Cochabamba, even though the altitude is 2,600 m. A "vetiver coordination center" is being established by the Organization for Environmental Conservation (AMBA).[21]

Brazil

Brazil produces vetiver oil for its own internal markets and has probably been growing the grass for several centuries. Nothing specific on its use in erosion control has yet been reported, but a scientific paper[22] reviewing vetiver in Brazil reported:

> It is a plant of great utility. Its numerous and tangled roots bind the surface where there is danger of breaking the earth apart; its tufts of erect, perennial leaves serve as a fence, protecting crops against wind and dust storms. Later its collected leaves, which have practically no odor, can be used to make hats and its straw canes serve to cover cabanas and barns. Its roots, which have a strong odor, can be transformed into baskets or coarse mats that in certain regions are suspended in doors and windows and frequently wetted with water, perfuming the air and lowering the ambient temperature.

CENTRAL AMERICA

Vetiver can be found scattered throughout Central America, but few people recognize it or are aware of the job it does. Nonetheless, rows of vetiver are often seen where roadsides have been cut into hills.

Costa Rica

Even before the initiatives in India began to arouse worldwide interest, plantings were increasing in an area of small farms and mixed agriculture southwest of San Jose. As a border to prevent erosion, the farmers say vetiver is better than lemongrass, which they had previously

---

[21] Information from Mathieu Kuipers.
[22] Gottlieb and Iachan, 1951.

used. Vetiver, they assert, tolerates more adversity, is more resistant to stem borers, lasts longer, requires less care, and (because it grows erect) interferes less with field practices.

The grass is commonly used as a hedge plant in the Meseta Central. It is also planted along the top of embankments to prevent rocks and dirt falling onto the road. An Indian group prepares the roots and sells them along with other local products, such as hats and baskets that they fashion out of local plants.

Guatemala

Guatemala once exported vetiver oil to the world. Although the trade stopped years ago, the plant still survives throughout the country. In a few places, it is even used as an erosion barrier. Engineers, for instance, have used it to protect road cuttings from washing out, and coffee planters have long relied on it for soil conservation. This has been mainly on plantations in the western coastal zones, especially in the Department of San Marcos.[23]

Some people (notably Indians in the hills) use vetiver leaves for thatch, for mulch to "break the rain" on seedbeds, and for bedding for pigs.

Recently, interest in vetiver has been renewed. The government and a private voluntary organization have formed a committee to coordinate vetiver promotion and research. Promising accessions have been located and were planted in six locations in October 1990. Only half of the plantings were watered, but, even though the dry season was just beginning, nearly all the plants survived.

NORTH AMERICA

Vetiver, of course, is a tropical species and would not be expected to grow in the temperate zones. Nonetheless, there are certain parts of North America where it survives and even thrives.

United States

Vetiver has been in Louisiana for at least 150 years. The roots were once routinely relied on to keep moths out of closets during the summers (for which purpose, it is said, the roots remained effective for two years). There was also a small industry producing vetiver oil.

However, this plant, once so well known to Southerners, has been

[23] The plant works so well that most people do not even think about erosion. A coffee planter, recently asked about how well it stopped erosion, replied "Gee, we don't know; we haven't had to go out there to see."

essentially forgotten since at least last century. Nonetheless, field observations suggest that in all that time the neglected plants have not spread, but have remained where planted. Today, vetiver can be found along the banks of many bayous and on old plantations. Even where homesteads were razed or abandoned during the Civil War, vetiver still grows.

The plant is known in other parts of the deep South as well. Although present in Florida for probably a century or more, it has never been collected as an escape from cultivation.[24] There was once a small vetiver-oil industry in Texas, centered mainly in the area near Riviera. The plant was also grown along the Gulf Coast as well as in southern California before World War II.

Recently in Louisiana, exploratory trials using the plant as an erosion barrier have shown remarkable promise (see next chapter).

OCEANIA

As related earlier, it was experiences in Fiji that stimulated the current rebirth of interest in vetiver. However, the grass is known to occur in many parts of the Pacific basin.

Australia

Several native species of *Vetiveria* grow in Australia (see Appendix B). All are found in the northern half of the continent. Recently, vetiver (the Indian species) has been tested with great success for its erosion-control abilities on highly erodible gullies near Brisbane.[25]

Fiji

The plant was apparently introduced to Fiji in 1907 and tested as a commercial crop before being let loose. Today, it is common in most parts of the islands and is sometimes used to bind rice bunds. Some people drink a tea made by boiling vetiver root in water. In some areas it has spread to populate roadsides and wasteplaces.[26] It is not, however, considered a threat.

Its use in erosion control in the 1950s (as described in the introduction) has been largely forgotten. However, because of newly mandated requirements of the Environmental Protection Authority, a construction company recently used it to restrain erosion and runoff from a building site.

[24] Information from Bruce Hansen, Herbarium, University of South Florida.
[25] Information from P. Truong.
[26] Parham, 1955.

Rakiraki, Fiji. Since planting his lines of vetiver grass 35 years ago, this farmer has successfully employed slopes that were formerly too dry and too unstable for sugarcane. The rows of vetiver take up space but otherwise interfere little with farming operations. In fact, they act as permanent guidelines that force animals (and farmers) to plow on the contour. That not only fosters soil conservation, it also increases the amount of moisture available to the crop plants throughout the field. (J. Greenfield)

Other Pacific Locations

American Samoa In Aunu'u, the plant has been used around taro fields to choke out weeds.

New Zealand Some samples have been introduced to New Zealand, but so recently that no results have yet been recorded.

New Caledonia In New Caledonia, vetiver has been used extensively to prevent erosion on slopes, particularly along roads. It proved notably effective.[27]

---

[27] Information from J. Barrau.

Cook Islands Plants have been on the island of Atiu for at least 30 years, and probably much longer. They show no sign of any natural spreading, and in some former locations they can no longer be found. One particular plant, in a domestic garden, is known to be 28 years old but is still less than 1 m across.[28]

EUROPE

Vetiver, of course, is a tropical plant and should not be expected to occur in Europe. There are, however, several exceptions: one is France.

Vetiver was introduced to the south of France as a potential source of ingredients for the perfume industry of Grasse, on the Côte d'Azur.[29] It still exists there and survives the Mediterranean winter.

Recently, it has been tested as a barrier against soil loss, both there and on the slopes of the Massif Central.[30]

---

[28] Information from D.E.K. Miller.
[29] Information from J. Barrau.
[30] Information from F. Dinger.

2

Case Studies

In some countries, vetiver was developed or tested as an erosion control completely independent of the World Bank efforts in Asia. These experiences add new insights into the merits and wider applicability of those described in the introduction. Moreover, they are in other parts of the world and in quite different types of sites. Thus, they form a complementary set of case histories that tend both to enhance and qualify those described earlier. Six examples are discussed below

UNITED STATES[1]

In 1989 Fort Polk faced a problem. This army base in Louisiana ("home of the Fighting Fifth Infantry, Motorized") is located at the headwaters of three scenic streams, which were filling with silt as tanks on training maneuvers ripped up the land.

The desecration of these pristine waters not only raised the ire of local communities, it threatened to bring down the heavy hand of civilian authority. Accordingly, army engineers laid check dams across the streams. That, however, did not solve the problem: after the numerous summer thunderstorms, turbid waters sluiced right over the top and muddied the streams as much as before.

Then Mike Materne, the local U.S. Soil Conservation Service agent, was brought in. By coincidence, he had just heard about vetiver. With little hope that it would do much good, he obtained some plants. (Possibly, they were remnants of some grown on Louisiana plantations before the Civil War and were still surviving, despite a century of neglect.)

His pessimism was all the more justified because the sites to be protected seemed hostile to any vegetation: the soil was down to almost bedrock ("C horizon"), the little that remained was very acidic (pH 4.0–4.2), and it contained virtually no fertility. But Materne

[1] Information based on the panel's visit to Fort Polk, September 1991.

decided to give vetiver a try—it would be more of a survival test than an erosion-control experiment, he thought.

Accordingly, Materne grew vetiver slips in large pots in a greenhouse, and in the spring of 1990 planted them side by side on the bare and barren slopes immediately above each check dam. He wanted to establish the hedges quickly, and so he dropped a tablet of slow-release fertilizer beside each plant. His hope was that any hedges that formed might filter the turbid waters and thereby stop dirt from ever defiling the dams.

Because of the siting of the dams, some of the vetivers had to be planted into waterlogged soil (owing to a recent downpour, they were standing in water when Materne left the site). Others had to be planted into pure sand, described by Materne as "drier than popcorn." To make matters worse, a gully-washer barrelled through before the plants had a chance to establish deep roots. The speeding water knocked out some and scoured out the soil around others.

Despite all these hazards, however, most of the plants in each of the four sites, from the wettest to the driest, survived. Moreover, a few of them withstood yet another adversity when a freak fire swept through one of the plantings. It scorched and even killed surrounding pine trees, but the vetivers all survived.

In fact they did more than survive—they thrived. In 8 weeks some were almost 2 m tall. In 10 weeks they had grown together into hedges. And on one site, more than 20 cm of sediment had built up behind the thin green line of grass. Some plants held back such a load that after a storm they were temporarily bent over, almost hidden beneath a "sandbar."

By that time the hedges were so effectively filtering the runoff that the old flow of mud and silt down the streams was largely cut off. The check dams were receiving mostly clear water and were functioning as designed: temporarily holding back surplus runoff for later release into the streams.

When first hearing of Materne's proposal to plant vetiver, the local county agent vehemently disapproved, arguing that introducing an exotic plant to the watersheds might create an uncontrollable weed problem. But he was mollified—even overjoyed—when native grasses, wildflowers, shrubs, trees, and vines came crowding in behind the hedges and grew to revegetate the site. He even declared that nothing like it had been seen in the area before.

By that time it was clear that vetiver was acting as much more than an erosion trap; it was a "nurse plant" that was protecting other species and thereby giving these devastated watersheds a chance to heal themselves. Whether because of better soil moisture or the captured silt, the combination of hedges and revegetated slopes solved what had seemed an intractable erosion problem little more than a year before.

Fort Polk, Louisiana. Tanks on training maneuvers tore up, compacted, and denuded the soil so much on parts of this army base that plants would not grow and the erosion seemed unstoppable. In 1990 and 1991, however, Mike Materne (shown here) planted vetiver grass across the worst of the washes. In this environment, so hostile to plant life, the vetiver hedges developed rapidly, trapping sediment even while they were still immature and full of gaps. (Twigs, leaves, and other debris carried by the water tend to lodge between the grass plants and helped plug the holes.) Within 10 months in some spots, sediments as deep as half a meter had built up behind these vetiver rows. (N. Vietmeyer)

ST. LUCIA[2]

Tall, lush, and rising abruptly from the sea, St. Lucia is a prominent island of the Windward chain in the West Indies. Vetiver has been there for perhaps a century and has been reducing soil loss on the volcanic slopes for at least half a century.

Today, vetiver is seen almost everywhere. As one St. Lucian explains, "Around here, people take on to it. The first thing they do when opening a new farm plot is to plant khus [vetiver] along the tracks leading to the plot."

Much of the vetiver seen today results from government interest in the past. From the late 1920s to the 1940s, for example, the government encouraged people to plant the grass. There was no formal policy, but the colonial agent would commonly push for it when he met with large landowners to play a little poker and discuss the latest farming techniques. Farmers on the periphery of the big estates imitated the large landowners, and thus the technique spread. To help the process along, several vetiver-demonstration plots were established at Mon Repos.

Today, the grass is sometimes found planted along the contours of hill slopes, where it functions much as the World Bank describes in India and Fiji. But more often, it is planted along the lower side of the swales (locally known as "drains" or "trained gullies"), which are ditches designed to carry excess water safely off the slopes. There, it reinforces the dirt walls to stop rushing runoff from bursting through.[3] Householders also use it to prevent mud and water from invading their backyards.

Although vetiver is well behaved and much sturdier than other grasses, hedges that are not maintained are said to "deteriorate." As one St. Lucian notes: "The unruly portions must be trimmed and the discipline maintained."

For instance, if the edges aren't cut back, the hedge may become ragged, perhaps because a few plants are unusually rambunctious or because the soil is of uneven quality. Some hedges may break up into clumps. Also, in some old and neglected plants, the centers die out. A timely topping helps keep them "tight."

The plant is easy to trim or top. Surplus vegetation is normally cut off with a shovel or machete. However, vetiver's roots loosen the soil so well that cutting back a hedge on the downslope side can expose highly erodible dirt to the elements.

---

[2] Information from Gabriel Charles, Standley Mullings, and Edithe Alexander.

[3] It is more effective to put it on both sides of the drains, especially on very steep slopes. It has been found that the best way to create these drains and ditches is to plant two lines of vetiver and to dig the trench between them only after they have become established. That minimizes both erosion and the damage from gully-washing storms.

Nowhere on St. Lucia are the plants considered weeds, and seldom do they spread beyond the hedges. Seedlings are never seen, although newly planted slips sometimes wash out and establish themselves down the slope, where they may look like errant seedlings. By and large the hedges seem to have little effect on neighboring crops, but plants immediately next to an old vetiver hedge sometimes exhibit a reduction in growth.

Another minor problem is that a small shrub—locally called "ti-baume" (*Croton astroites*)—can overcrowd older hedges. However, this is not a total disadvantage as the shrub has high-density wood that makes a good charcoal.

One hazard to which the hedges are immune is St. Lucia's fearsome feral goats. These animals are so destructive that people say "all goats' mouths are poisonous." The problem is especially bad at the end of the dry season, when grasses have been grazed out and the goats have started on the trees. The animals, however, graze around vetiver, and the erosion control is unaffected.

In recent decades it is not goats but people who have destroyed many of St. Lucia's vetiver hedges. "Now everyone wants to be very modern and build a wall," says one disgusted observer. "The trouble is, it's more expensive and less effective."

Another resident notes that: "Years ago—when khus was everywhere and the slopes were forested—rivers ran year-round, but now in places they aren't running at all in the dry season because the soil is gone. People are building up on the slopes, and there is more and more water charging down in the rainy season. Everything gets mucky, but everyone wants a house!"

More vetiver, it seems, might be the solution—as it was in the past.

INDIA[4]

In 1956, the National Botanic Gardens (NBG)[5] in Lucknow initiated what seemed a pointless endeavor: a major effort to reclaim a patch of usar soils. These soils, which cover nearly 7 million hectares of India, are so alkaline and salty that they have long been classified as unfit for agriculture.

However, the director of the NBG, K.N. Kaul, decided to tackle the impossible. He began his project around the village of Banthra (just outside Lucknow, on the Kanpur road), where many hectares of usar soil had been lying unused for years.

---

[4] This section is based on information from T.N. Khoshoo. Further details can be found in T.N. Khoshoo, editor. 1987. *Ecodevelopment of Alkaline Land: Banthra—A Case Study*. National Botanical Research Institute, CSIR, Lucknow, India.

[5] Renamed the National Botanical Research Institute in 1978.

To those who saw the area, the prospect of producing anything useful appeared bleak. The soil was bare, hard, and highly eroded. Hard pans had surfaced in places, and a thick crust of sodium clay stretched as far as the eye could see. The alkalinity was extreme (as high as pH 11) and just 1 m below the surface was an impermeable layer of calcium carbonate that blocked plant roots and produced widespread waterlogging in the monsoon season. The only vegetation to be seen was sparse clumps of grasses and isolated specimens of the weed *Calotropis procera* with its salt-filled bladders and toxic leaves.

Administrators from the state government felt that this experiment, like all the previous ones on usar soils, was bound to fail.[6] They gladly made the site available without charge. After all, what had they to lose? The people of Banthra were living in utter poverty, and many had resorted to crime to survive. Even the Banthra people themselves were less than enthusiastic, convinced that cultivating such hard and barren land would demand tractors, bulldozers, subsoilers, rollers, and other heavy machinery. They could foresee only big costs and small rewards.

However, they soon found that things were to be different. Professor Kaul planned to employ not machinery, but organic amendments and natural methods. His goal was to create a self-sustaining agriculture based on alkali-tolerant herbs, shrubs, and trees. It seemed a good idea, but at first nothing would grow at Banthra. Even the most resilient food crops died of stress and exposure. This project certainly seemed to be the failure everyone had predicted.

But then everything changed, and it was vetiver that made the difference. This rugged grass possessed an exceptional ability to withstand the heat, the drought, the salt, the alkalinity, and the waterlogging. Without even amendments or fertilizers, it could establish itself when planted directly into the usar concretion.

And vetiver did much more. As was later discovered in Louisiana, it proved to be a "first aid" plant that started the process of healing the site. Vetiver stretching in rows across the land gave other plants a chance to survive, too. It blocked the drying winds and reduced the erosion they caused. Indeed, the better microclimate and environment between the rows helped the NBG researchers establish a workable farming system. In the process, the soil began slowly to improve. This

---

[6] Their skepticism seemed justified: prospects for working with such land were bleak. Iron nodules were prominent in the soil profile, drainage was poor, alkalization was severe, the water table was at 4–5 m below the surface, pH was 8.5–11.0, electric conductivity (at 15 cm depth) was 0.7 m mhos per cm, and exchangeable sodium ranged between 40 and 73 percent. In addition, the average organic carbon was only 0.2 percent, available phosphorus 7.9 kg per hectare, and potassium 300 kg per hectare.

Through vetiver, this . . .

Near Lucknow, Uttar Pradesh, India. Even the researchers involved in the project find it hard to credit the transformation that occurred at the village of Banthra. The picture above shows the scene before they began trying to revegetate the site in the mid-1950s (see text for more details). The area was then a barren and unusable wasteland, toxic to almost all vegetation. The picture on the right shows the same scene in the mid-1970s.

. . . became this

Vetiver made this possible. Its ability to survive the alkalinity and harshness of these degraded plains provided an initial plant cover that helped stabilize the land. The vetiver rows reduced the excessive effects of wind and sun and erosion, and other plants could then be established as well. In 20 years, a remarkably short time, this stable forest resulted. (T.N. Khoshoo)

was especially so after the researchers dug drainage systems and created ponds for collecting runoff and recharging groundwaters.

The land at Banthra is almost flat, and Professor Kaul initially thought of vetiver not as an erosion-control barrier but as a potential commercial crop. The villagers could sell the roots for essential oil, he thought. Nonetheless, the outcome exemplifies the plant's ability to survive adversity and to foster the growth of relatively less tolerant species.

That is certainly what happened at Banthra. Today it is a lovely parkland: green, shady, and beautiful. Legumes of various types now flourish, and 18 species of plants (belonging to 15 families) that had not been recorded there at the time Professor Kaul began his work are now common. The land has been transformed. It now supports a healthy mix of woodland, grassland, and cropland. As the leader of the project at the time stated with relief, "The final proof came with the accumulation of humus and eventually with the appearance of earthworms. Although this took 12–15 years, it was a good reward for the efforts we had started in the mid-'50s with 'first-aid' species like vetiver. This was indeed a day of rejoicing for us all."

MALAYSIA[7]

In Malaysia, where the rain can fall in sheets and the slopes are steep and loose, soil losses can be among the severest in the world. Vetiver hedges would seem to be a godsend, but (at least in recent times) they were unappreciated until P.K. Yoon read about the World Bank's results in India and immediately set out to see for himself if vetiver worked.

Yoon is a scientist with the Rubber Research Institute of Malaysia, and he embarked on this venture in 1989 with less than great enthusiasm: "When I first saw a clump of rather undistinguished-looking grass, it looked so ordinary and so frighteningly similar to the horrible 'lalang' [*Imperata cylindrica*, a feared tropical weed] that I was thoroughly put off. However, also having seen massive erosion problems, especially on steep hills, I was prepared to have a look-see at anything that might work."

Luckily, Yoon managed to locate a vetiver clump near the city of Taiping. He carefully broke up the clump into 57 separate plants (tillers) and planted them out in individual polybags.

Vetiver proved easy to multiply. Much watering and a little slow-release fertilizer greatly boosted the growth and the production of

[7] Information from P.K. Yoon. More details can be found in P.K. Yoon. 1991. "A Look-See at Vetiver Grass in Malaysia: First Progress Report" (see Selected Readings).

tillers, and topping the clumps back to 40 cm encouraged tillering even more.

At first, Yoon threw away the tops that had been cut off; however, he eventually noticed that as long as the plants were at least 3 months old, the discards included many culms. These jointed stems had buds at each joint, and Yoon found that laying the stems on damp sand and keeping them under mist caused the buds to sprout and produce new plantlets. Slitting the leaf sheath increased the success rate to the point that, after just 8 weeks, three out of every four nodes took root and began to grow. With as many as 17 nodes to a culm, this has proven a quick and efficient way to propagate vetiver without ever digging any up.

By these methods Yoon was able to convert the 57 starting plants into an amazing 16,000 tillers in just the first 7 months. Within 18 months he had distributed 200,000 plants for testing at various sites throughout Malaysia.

To see if vetiver had any chance of stopping soil loss, Yoon set up a simple demonstration at his research station at Sungei Buloh. The terrain was gently undulating (4–5° slope), and erosion had already formed a small gully. He planted slips of vetiver across the gully in five widely separated rows. After only 3 months, they had grown into hedgerows and had trapped so much topsoil that the gully had gone; in fact, what was previously a gentle slope had become level platforms, each faced by a bristly line of grass (see top of next page).

As a demonstration, it was very successful. "Every visitor we've had to date has been impressed enough to want to use vetiver on their own land," notes Yoon.

Next he set up a larger demonstration. A hillside was divided into four portions; two were planted with vetiver, the other two with cow grass (*Paspalum conjugatum*) and New Guinea grass (*Panicum maximum*) as is normally practiced in Malaysia. At the side and bottom, corrugated-metal walls were built to deflect runoff into drums, where it could be measured and its silt content analyzed. All this was rendered useless, however, when such heavy rains fell for 2 days in October 1990 that the runoff collapsed the metal walls and washed away the drums. The vetiver-planted portions remained pretty much intact; the rows of vetiver did their job, but with no control slope for comparison the precise figures Yoon had hoped for could not be obtained.

Yoon's other work concentrated on testing vetiver's ability to protect highway embankments, steep banks in housing estates, and hillsides in large new plantations. In such sites, saving a few dollars in propagation and planting costs is trivial, and Yoon grows out the plants in polybags to ensure that they rapidly produce uniform hedges when placed out on the site.

Vetiver in Malaysia

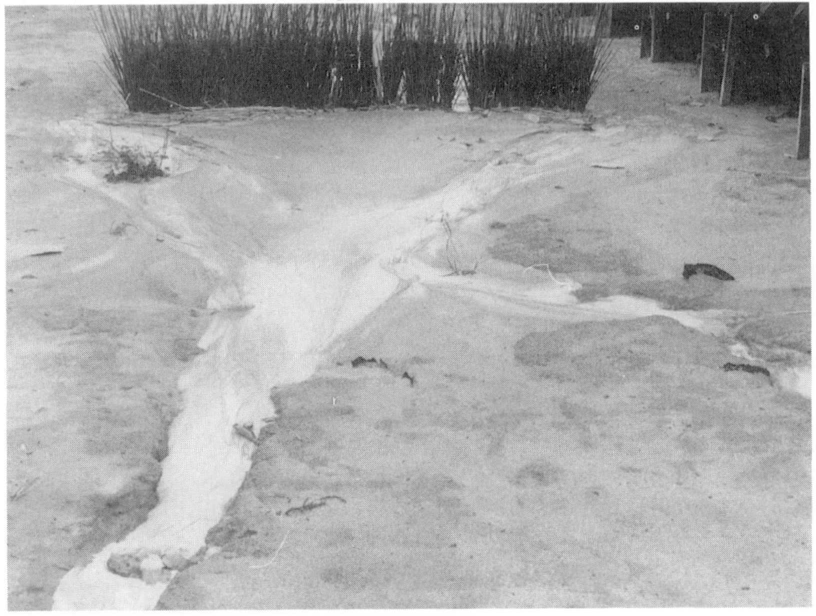

Under ideal conditions vetiver hedges can trap topsoil rapidly. The photograph shows a site merely 2 months after a small hedge was planted across a fast-developing washout. (P.K. Yoon)

Rows of vetiver were able to stabilize fill dirt within a few weeks of planting. Although not really established in the loose and highly erodible soil, they protected the site during one of the severest downpours in memory. The bare section adjacent washed out. (P.K. Yoon)

Vetiver stabilizing the edge of a hillside road, a common location for severe erosion. (P. Boyer)

Vetiver hedgerows protecting newly formed embankments around a deep pond. This plant, which withstands wet conditions and even total immersion, seems an excellent candidate for protecting the edges of waterways of all types. (P.K. Yoon)

Concerned that the grass might host diseases that could affect Malaysia's crops, Yoon has surveyed the various plantings for pests. However, despite the large number of sites, there were only two serious fungal attacks—both of them in crowded nurseries.[8] Moreover, the problem was easily solved: topping the plants removed the diseased parts, and the subsequent growth was normal.

An orchardist who had received some of Yoon's plants discovered an interesting application. To provide the abundant water essential to his new orchard, he excavated a series of ponds and planted the grass on the embankments. The plants grew extremely well, and they stabilized even the filled-earth sections so quickly that within 5 months the embankments were able to hold back waters 3 m deep. During the rainy weather many of the plants survived more than a month under water.

The key innovation, however, occurred when the orchardist added fish to the ponds. "He found that Chinese grass carp love to eat vetiver leaves," notes Yoon, "and he now cuts off the top of the grass and routinely feeds it to the fish. He is so pleased with this that he is digging more ponds. Also, neighboring smallholders are introducing vetiver and fish-rearing to their farms."

Young tops were readily consumed by sheep as well. Moreover, repeated cutting not only feeds the animals, it keeps the hedgerows dense and neat and in the finest condition for controlling erosion.

Yoon has also tried vetiver tops as a mulch. Farmers and gardeners in Malaysia commonly use lalang for this purpose. Preliminary observations showed that vetiver lasted much longer and did not cause any weed problem. Analysis showed that the vetiver mulch had considerable amounts of available nutrients (including nitrogen, phosphorus, potassium, and magnesium). Yoon concluded, therefore, that vetiver will be used for much more than just erosion control; it will also be a ready supply of quality mulch for suppressing weeds, conserving soil moisture, and boosting crop yields.

All these hands-on experiences have converted the former skeptic into a vetiver enthusiast: "Our work was done over a period of less than 2 years, but the results clearly show vetiver's vast potentials. They are just too tempting for anyone *not* to look further into it!"

SOUTH AFRICA

Vetiver has been part of Tony Tantum's life since 1966, when he first discovered Mauritian sugarcane farmers using it to stabilize drains

[8] Three different fungi were identified as species of *Nigrospora*, *Curvularia*, and *Helminthosporium*.

in Malawi. To him, the hedges on that sugar estate in the Lower Shire Valley seemed very effective. At the time he learned that vetiver also had other uses: the culms were used as brooms, and small bundles of the roots, tied with a ribbon, were used for scenting cupboards.

In the 1980s, he moved to South Africa and found that the South African sugar industry had been using vetiver for a long time—perhaps for more than 100 years. "Hedges of the grass were put in to keep equipment from falling off the slopes," explains Tantum. "However, this was not practiced by everyone; it was pretty much limited to the French Mauritians growing sugar in Natal. They don't get along with the other sugar growers, so the technique did not spread."

Today, however, this method of erosion control is escalating. Indeed, Tantum has built a broad national base of institutional awareness of vetiver. His main resources have been Cedara College, Elsonberg College, the Institute of Commercial Forestry Research, the Environmental Authority, Natural Resources, and administrators in Transkei, QwaQwa, Bophuthatswana, and KwaNdebele. He has, moreover, generated a similar network of institutional interest in Lesotho, Zimbabwe, and Mozambique.

Tantum reports that he meets with skepticism and resistance from mechanical engineers (which is not surprising since engineered systems have dominated South Africa's erosion control for 30 years) but that most people are persuaded pretty quickly.

One convert is a North Coast farmer who was on his ninth ratoon (annual cutting) of sugarcane and preparing to replant the field when Tantum persuaded him to put in vetiver hedges. The grass would act as an "erosion safety net" when the cane was plowed out the following year, he said. But the unexpected occurred: the ninth ratoon turned out to be the farmer's highest yielding field. Perhaps the dense grass hedges were improving soil moisture to the point where an unprecedented tenth ratoon would be profitable; perhaps not. But the farmer faced a dilemma: keep the cane or replace it?

As a result of such experiences, the South African Sugar Experimental Farm is conducting a 5-hectare study of soil loss and water retention in sugarcane using vetiver.

Tantum's company, which puts in vetiver hedges under contract, has helped establish vetiver trials all over the Republic. Several are on horrendous sites. In the Western Cape, for example, the grass has been established on pure sand and on badly eroded kaolin clay. Elsewhere in the Cape, on dumps of slime residue from cement manufacture, it is doing well despite the worst drought in many years.

The trial on the pure sand was particularly impressive. It was on the coast near Camps Bay, on the road to Llandudno. Eroding road banks have long been a problem in the area. When Tantum's crew arrived, the debris of many previous attempts—poles, wire, and plastic netting, for example—was evident. It was clear that this would be an extreme test for vetiver. The soil was barren orange grit and the steep slope faced the sea, exposed to salt spray and sea winds. The grass was planted in April 1990, and more than 90 percent of the plants survived. Within about 2 months, they started to stabilize the site.

Tantum has also established trials for several government agencies. The Department of Railways, for instance, approached him when a steep (1:1) embankment near Shongweni collapsed after a downpour of 100 mm. The embankment had supposedly been stabilized with a covering of kikuyu grass. The railways department asked Tantum if vetiver could do better. Tantum planted rows of vetiver straight into the collapsed, eroded, and unprepared surface. The soil was very poor, but no fertilizer was used. Within one month the rilling had stopped. Within 2 months residual kikuyu grass had begun to cover and stabilize the soil between the vetiver lines. However, neighboring areas (lacking vetiver) remained bare. For them, soil had to be imported and the area resodded with kikuyu grass.

In 1990, Niels Carstens of the Roads Department (Cape Provincial Administration) asked Tantum if vetiver could solve the serious erosion at the Stellenbosch flyover (Exit 22 of Highway N2 to Cape Town). The embankment here was very steep; the so-called "soil," pure white sand. Vetiver was planted in April 1990, closely spaced and without fertilizer. Virtually all the plants survived, and natural terrace formation was already visible before the end of the year.

In another trial in the Stellenbosch area (on the R44 road to Paarl), a steep road bank with very poor white clay subsoil was planted. Nothing grew there until April 1990, when vetiver was put in. Within 7 months the grass was tillering well, and the bottom hedge had built up between 70 and 100 mm of soil.

An interesting project, developed by the Institute of Commercial Forestry Research, has used vetiver hedges to stop soil loss in firebreaks, which were a major source of erosion. Thanks to the institute's work, the South African forestry industry has now accepted vetiver for this use. Also, the insurers of the industry have accepted that vetiver hedges on firebreaks are not a fire hazard. In June of each year, the hedges are treated with a contact herbicide and burned a week later. Within 2 weeks they become green belts across the firebreaks, blocking the former erosion.

Chris Nicolson, of the Institute of Commercial Forestry Research, is now developing an evergreen ground cover to fill the space between the hedges. This is a major development for the industry, and perhaps not only in South Africa because firebreaks are a source of erosion in forests worldwide.

MADAGASCAR

It is no news to anyone that Madagascar has a problem with soil erosion as bad as can be found anywhere in the world. On the cultivated uplands (slopes up to 100 percent or more), minor surface rilling rapidly evolves into fierce gullying that gouges out vast areas and turns the rivers to soup. Often, this gullying starts along the elaborate networks of drainage ditches the farmers dig around their fields. These ditches are intended to carry away the runoff, but they often wash out. Even when intact, they take away valuable moisture so that the mountain soils tend to quickly dry out when the rains cease.

Currently, however, the only other efforts to halt this disaster on Madagascar consist mainly of scattered reforestation projects. Their impact is often minimal because, in the absence of adequate ground cover, few of the seedlings survive and those that do grow slowly. Moreover, many end up destroyed by bushfires that recur each year.

All this was known to Thomas Bredero, the World Bank's senior agriculturist in Antananarivo. Thus, in 1988, after seeing how well vetiver performed in World Bank projects in India, he began searching Madagascar for the grass. Fortunately, French colonists had previously introduced it to produce vetiver oil. Bredero found remains of their plantations, although they were scattered and few. His main problem was how to finance and establish nurseries and demonstrations throughout the country. His first effort failed because the grower demanded gold for every vetiver plant he produced. A second was more successful, despite a seemingly never-ending drought that began just after planting. This time, a commercial farmer produced about 10 hectares of nursery, and about 90 primary schools in the Lac Alaotra area planted nurseries of 1 hectare each. As a result, farmer groups were soon being provided with vetiver slips. By late 1991, on-farm demonstrations could be found in 11 of Madagascar's 22 extension districts. It was at this point that Bredero came to be known as "Monsieur Vetivère."

However, it was clear from the outset that getting farmers to accept vetiver was not going to be easy. "There were the usual arguments—that it has no 'economic' purposes (such as improving soil fertility and cattle fodder)," said Bredero, "and that other 'economically more useful' species are available."[9]

But much of the opposition subsided when Bredero planted a vetiver hedge in the presidential garden. It solved an incipient erosion problem on the palace grounds and greatly pleased the president.

---

[9] These included species of *Setaria* and *Pennisetum*, but, as yet, they have not been developed to any large extent for vegetative contour protection.

Madagascar's extension service now recommends vetiver for on-farm soil and water conservation in combination with other measures such as contour cultivation, dead furrows, continuous vegetative cover, and crop rotations. On slopes under 5 percent, where burning is not practiced, a grass called "kisosi" (a species of *Panicum*, see Appendix B) is also recommended.

In these combinations, vetiver is employed as a first line of protection, not only against erosion but also against ground fires. It complements agriculture, horticulture, and reforestation. With well-established vetiver lines, for example, many other kinds of land uses that lead to soil conservation are being developed: annual crops, perennial crops (notably fruits and fodders), and reforestation, for example.

Although at first skeptical, the forestry research department[10] is now supportive. It changed its position when tests on its own sites showed that vetiver by itself slows runoff as well as a dense forest cover could.

In a number of Madagascar's rural areas, farmers have discovered for themselves vetiver's effectiveness for stabilizing dams, rice-field bunds, and irrigation works, as well as for protecting roads that can flood and wash out.

Bredero's next major challenge is to prove vetiver's usefulness in preventing the devastating gullies and ravines (known as *lavaka* in the Malgache language) from chewing up more land. They are so big and there are so many of them that the sandstone formations north of the capital and around Lac Alaotra constitute an alarming sight.

Bredero is now tackling the problem from two sides. First, vetiver is planted on contour lines around the upper edges as well as down the sides of the ravines to slow down and disperse runoff coming from the top of the mountains. Second, wooden poles are driven into the sand at the bottom of the ravines. The soil retained by these wooden palisades is planted with vetiver, bamboo, and fast-growing and fire-resistant trees and shrubs.[11] The result is a dense vegetative cover. About 10 of these pilot-sized watershed-protection projects are now established, and early experiences seem encouraging.

The final verdict on vetiver is not yet in, but this grass just might be the answer to Madagascar's raging erosion—one of the worst local environmental problems on the planet.

---

[10] Of the national agricultural research organization, FOFIFA.

[11] Species such as *Acacia dealbata*, *Grevillea banksii*, *Tephrosia* spp., and *Calliandra calothyrsus* are employed.

3 Conclusions

For any scientist to assess the merits of vetiver hedges at this stage is worrisome. Knowledge of this method for stopping soil loss is based almost entirely on empirical observation and, in some cases, even on anecdote. Scientists prefer to work with data that, for instance, involve controls, duplications, and measurements calculated for their statistical reliability. With vetiver, few such figures or factual comparisons are available.

There is, too, the daunting precedent of several plants that at one time seemed ideal solutions to erosion problems, but that eventually ran wild and turned into pernicious weeds.[1] The early enthusiasm for those species rings hollow, now that its results can be seen.

Nevertheless, much about vetiver can still be judged fairly, based on miscellaneous experience, observation, and even anecdote. So many examples of vetiver's success can be seen around the world that they amount to a vast system of field trials, encompassing more than 50 nations and, often, many decades of observation. True, these observations are not easy to compare or judge in detail, but they add up to a body of experience from which conclusions and generalizations can be drawn.

Further, it is important also to place this in context. In global terms, erosion is continually increasing. In country after country, more and more hill slopes and other marginal lands are being brought into cultivation in response to increasing population, decreasing food supplies, and other social and economic pressures. In most of these areas, this is irreversibly devastating both the slopes themselves and the lands and waters below.

---

[1] These, however, were spreading plants used to cover and clamp down wide expanses of exposed soils. Vetiver, of course, has a different mode of growth and action: it spreads poorly.

Fort Polk, Louisiana. A vetiver hedge can accumulate an amazing amount of soil. Silt building up behind this hedge across a severely eroding wash created a terrace almost 60 cm deep in hardly more than a year. (N. Vietmeyer)

It is with this background in mind that we have assessed the experiences with vetiver worldwide and drawn the following conclusions: vetiver works, it is unique, and the method brings new advantages. Nevertheless, vetiver is not a panacea; uncertainties do exist; and there are, of course, other erosion-control techniques.

Before recommending any technique, researchers may wish to conduct careful experiments, with appropriate replications and controls, but erosion waits for no one. Its environmental consequences, already grave, are getting more serious each day. Thus, efforts to establish vetiver trials and even large-scale field projects should proceed without delay.

VETIVER WORKS

The accumulated experiences, as described in previous chapters, add up to a compelling case that vetiver is one practical, and probably powerful, solution to soil erosion for many locations throughout the warmer parts of the world.

To the casual observer, it may seem implausible that a hedge of grass only one plant wide could block the movement of soil under torrential tropical rainfall. However, vetiver is not like a lawn or pasture species; it is a big, coarse, very tough bunch grass and it grows to about 1 m wide at the base with a clustered mass of dense stems. Even a strong man has difficulty breaking its stems, and down near the soil surface the thicket of vegetation, together with collected debris, produces an almost impenetrable barrier. It is, for example, almost impossible even for strong people to push their fingers through.

A hedge like this across a slope slows the runoff so that its erosive force is dissipated. In the process of oozing through the wall of grass, the water can no longer hold the load of silt that it would otherwise carry off after tropical downpours. Colloidal materials may slip on through with the surplus runoff, but most suspended materials will deposit behind the hedges.

Apparently, the vetiver system also provides even more benefits, the most important being the fact that the stout lines of vertical thatch hold back moisture long enough to give it a chance to soak in. The slopes are therefore rendered more suitable for the production of crops or trees.

Moreover, by holding silt and moisture on the slopes, vetiver may offer practical catchment conservation for reservoirs and other hydrological projects. In many places this will undoubtedly prove impractical, but it could nonetheless be invaluable to certain projects involving canals, rivers, reservoirs, flood-control facilities, and other waterways to even out the flow of runoff throughout the year. Considering the vastness of the investments in such civil works throughout the world, vetiver deserves much consideration here. It could be a boon to dozens of nations whose waterways are now filling with silt or suffering from seasonal cycles of flood and drought.

VETIVER IS UNIQUE

Although other grasses and trees have been used as vegetative barriers for soil conservation, vetiver seems to combine several characteristics that make it special:

- It reduces erosion when in a hedge just one plant wide. (Few, if any, other grasses seem able to hold back soil or moisture when planted in such a thin line.)
- Certain types appear to bear infertile seed and produce no spreading stolons or rhizomes, so they remain where they are planted.
- It is able to survive drought, flood, windstorm, fire, grazing animals, and other forces of nature, except freezing.
- It has a deep-penetrating root system.
- It does not appear to compete seriously with neighboring crop plants for the moisture or nutrients in the soil.
- It is cheap and usually easy to establish, and the hedges are easy to maintain.
- It is not difficult to remove if no longer wanted.
- It is (at least so far) largely free of insects and diseases and does not appear to be a host for any serious pests or pathogens that attack crops.
- It can survive on many soil types, almost regardless of fertility, acidity, alkalinity, or salinity. (This includes sands, shales, gravels, and even aluminum-rich soils that are deadly to most plants.)
- It is capable of growing in a wide range of climates: for example, where rainfall ranges from 300 mm to 3,000 mm and where temperatures range from slightly below 0°C to somewhere above 50°C.

VETIVER BRINGS NEW ADVANTAGES

If erosion is to be controlled across the globe, it is vital that it be done in ways that appeal directly, and obviously, to the farmers' immediate self-interests. Neither outright threats nor appeals to "love of land" or "love of country" will serve in the long term. Ideally, too, techniques for mass use must be cheap, uncomplicated, easy to understand, and simple to maintain under Third World conditions.

Vetiver can be all of these things.

Oxford, Mississippi. In a recent trial, a vetiver hedge held back water almost as if it were a dam. This test was conducted in a flume (61 cm wide) and the water (flowing at 28 liters per second) was ponded to a depth of 30 cm behind the hedge. This result was all the more remarkable because the hedge was young and less than 15 cm thick. (Sedimentation Laboratory, Agricultural Research Service, U.S. Department of Agriculture)

Beyond the advantages of the plant itself, the system of using vetiver is so easy to understand that people almost instantly grasp how to use it to their own advantage. Moreover, putting in vetiver hedges has several other benefits:

- It does not require that each site be individually designed.
- It does not require foreign exchange or expensive equipment.
- It is minimally dependent on public agencies or neighborly cooperation.
- It does not need laborious maintenance.
- It does not require careful layouts or high-quality control.

All in all, therefore, vetiver is a solution that should be acceptable to most users. It seems promising as a way for local people to involve

Rakiraki, Fiji. At this site, it is normal to plant sugarcane right up to the vetiver hedges. Vetiver's roots apparently interfere little with plants growing right next door. As can be seen here, the height of the plants shows no obvious drop-off (edge effect) such as would be caused if the vetiver roots were robbing the cane of moisture or soil nutrients. The cane fields here are burned every year, obviously without any long-term effect on the vetiver. (D.E.K. Miller)

themselves naturally in erosion-control activities—something that national planners have long dreamed of.

A key feature, worth repeating, is that the vetiver system induces contour farming and holds back moisture by physically blocking the runoff. Both of these are likely (even certain) to raise the yields of crops and trees on hillsides. Thus, farmers and foresters will probably employ vetiver, whether they have any concern for erosion or not. Self-interest should drive them.

VETIVER IS NOT A PANACEA

Vetiver is not a panacea; it cannot solve all the problems. Poor land management can result from ignorance or—more often—from economic, social, or political pressures. For example, rents may be too high, fertilizers unavailable, or crop prices too low—all of which can force farmers to overexploit their land in their attempts to grow more.

Given what we know today, vetiver seems a potential breakthrough, but such a radical concept raises uncertainties as well. Some difficulties, however, can already be foreseen.

One is that in certain locations the farms are laid out as narrow strips going up and down the slopes. In practical terms, such farms can be plowed only vertically. Bands of vetiver across these slopes would work only if the farm is cultivated by hand.

Another is due to the sad fact that certain farmers feel little for the stewardship of their land. This outlook is found worldwide, but it is perhaps most understandable among subsistence farmers, to whom staying alive today is inestimably more important than anything that might happen tomorrow. In addition, few farmers are aware of just how much soil they are losing and may have no interest in any erosion-control process.

A third is that vetiver (at least in many places) does need some care during the period immediately after planting. Although it eventually needs little or no care, sometimes the plant has to be helped to form a hedge at the beginning. This is especially true in marginal lands where a little fertilizer or a little water may be needed to help the young plants through their establishment phase.

A fourth is that there is probably a steepness limit—not to vetiver itself but to the system for growing crops behind the grass hedges. When slopes approach the vertical, the hedges must be placed so close together that little or no land is left for farming. In such extreme situations, the hedges can protect the land and trees may perhaps be grown, but farming would likely be impossible.

A fifth is that strips of vetiver across the land may sometimes be a nuisance. (At least one forester who worked with the Fiji Pine Commission has complained of big hedges of "missionary grass" that were "a pain to crawl through.") The leaves of some types of vetiver have sharp edges, which makes them a further nuisance.

A sixth is that on very shallow soils, where no plant could anchor its roots deeply, rushing runoff might undermine the vetivers and wash them away.

Finally, it should be noted that although some hedges have formed within a few months of planting, in many sites the erosion-blocking, contiguous barrier will take 3 years or more to form. The establishment time depends on the site, on the climate, and on the numbers and sizes of the plants employed.

Vetiver may fail to form functional hedges on sites with only moderate temperatures and sunlight. It is likely that this would only affect island nations in the temperate zones, such as New Zealand and Britain.

Why Conservation Schemes Fail

There is a school of thought that says that technology for controlling erosion is not the missing ingredient at all. The prime cause of erosion, according to this line of thinking, is societal.

"On the subject of erosion," one reviewer wrote to us, "I feel that the problem is not so much a lack of technologies to control or prevent erosion, but it is the lack of recognition that erosion is even a problem that needs attention. Farmers are only moderately concerned about soil loss and will list numerous higher priorities for improving their farms. Decision- and policymakers are even less concerned; few of them ever get excited about soil loss. Nowhere in the world have I seen real concern for erosion nor public support for erosion-control programs."

Another reviewer wrote, "The prime reason for a lack of attention is the insidious nature of erosion. One rainstorm can remove a millimeter of soil from one hectare. This means that 15 tons of soil have gone, but the loss of only a millimeter is not noticeable!"

Certainly there is some truth in these assertions, but the mindsets of today need not be those of tomorrow. Changes can be made. Piers Blaikie,* for instance, claims that governments can create the right policy environment for attacking the causes of soil erosion by stimulating such actions as:

- Research (notably on specific techniques);
- Legislation (for example, banning very damaging practices);
- Supporting extension and facilitating credit (and tying both to the positive encouragement of soil-conserving crops);
- Rural development (for instance, strengthening local institutions);
- Improving administrative structures (so that bureaucratic decisions are based less on fiat and more on experience);
- Land tenure reform (so that farmers have a vested interest in erosion control);
- Adjusting prices for farm products and inputs (especially eliminating those that subsidize unsound practices); and

* P. Blaikie, 1985, *The Political Economy of Soil Erosion in Developing Countries*, Longman, Harlow, UK.

• Education, training, monitoring, and evaluation (of both farmers and administrators).

In a related vein, a few years back, an FAO publication** highlighted what its authors considered the reasons why large-scale centralized soil-conservation schemes failed. It identified the following reasons:

• Erosion prevention is seen as an end in itself.
• There are very high labor requirements.
• The effects on agricultural production are usually ignored.
• Farmers, seeing few short-term benefits, lack motivation.
• Farmers and herders are regarded as part of the problem to be solved.
• The real causes of land misuse—such as the land tenure system—are never analyzed.

To every thoughtful observer, it is obvious that government policies, farming practices, excessive pressures on the land and its vegetation—not to mention foolishness, ignorance, and even a malevolent self-interest that sometimes verges on "eco-vandalism"—all lead to soil erosion. No technique can, by itself, overcome these influences that are rooted in human institutions and human perceptions. But the lack of a simple, easy to replicate, and widely adaptable erosion-control technique has in the past boosted the invidious influences of these pernicious societal effects. Vetiver, therefore, might help. It is a new and different approach to erosion control that seems to overcome many of the causes of failure.

Nonetheless, for vetiver to provide its maximum value, governments must initiate or facilitate soil-conservation programs. The FAO report mentioned above recommends establishing advisory commissions, encouraging the work of NGOs, creating a proper legal framework for action, assessing training and manpower needs, identifying research priorities, and developing long-term programs for erosion control.

** Food and Agriculture Organization of the United Nations (FAO). 1990. *The Conservation and Rehabilitation of African Lands: An International Scheme*. Publication No. ARC/90/4. FAO, Rome.

VETIVER IS COMPATIBLE

Vetiver isn't the only erosion-control technique, of course. Others include the following:

- Engineered systems, such as terraces, rock walls, and earthen berms and bunds;
- Plants that spread over the land;
- Broad (as opposed to narrow) strips of grass;
- Tied ridges;
- Contour cultivation, mulches, crop rotations, strip-cropping, and no-till farming; and
- Forestry, agroforestry, and living fences.

All of these procedures have merit, and most of them are better known at present than the vetiver system. Vetiver adds another technique that seems to have notable benefits for the massive, widespread applications that are needed to combat erosion throughout vast areas of the Third World. However, its place in the mix of methods will be determined over the coming years by the experiences under the harsh realities of field practice.

Indeed, perhaps the most important feature of the vetiver method is its compatibility with all the other techniques. Vetiver is already being planted in several countries to reinforce and improve the stability of terraces, berms, and bunds. It has outstanding promise as a "safety line" to anchor broad strips of other grasses, such as napier grass. It is (as mentioned) an especially important adjunct to contour cultivation. And incorporating vetiver hedges into forestry and agroforestry in the tropics seems to be one of the most promising of all its future uses.

4

Questions and Answers

This chapter presents 22 of the concerns most commonly raised about vetiver, together with what seem to be the best available answers. These pages reflect the general state of knowledge in the early 1990s. They result from reviews and comments collected from a dozen experienced vetiver specialists and a dozen vetiver skeptics.

Will Vetiver Stop the Loss of Soil?

The accumulation of experiences worldwide is convincing: vetiver hedges can indeed block the passage of soil. They may not stop a hillside from slumping, but they can keep topsoil on site and, over time, retard most surface erosion. In many cases they can also help fill up gullies.

With the experiences of India, Fiji, China, Malaysia, St. Lucia, St. Vincent, South Africa, Malawi, the United States, Haiti, and other nations, it can be said with confidence that hedges of this deeply rooted perennial are indeed capable of catching eroded soil and helping to build up terraces that stabilize the site. Moreover, they seem capable of doing so under many conditions except where freezing occurs. The actual amount of erosion controlled on any given site cannot now be predicted with confidence, but, at least in some cases, the reductions can be dramatic.

It should be understood that, in a strict sense, vetiver doesn't stop erosion. The soil between the hedges is still free to move. However, the hedges block its further progress and prevent it from leaving the site.

Although it has the power to hold back soil, there are undoubtedly sites where vetiver will fail: climates may be too dry or too cold, slopes may be too steep or too boulder-strewn, or soils too toxic or too shallow, for example.

Is It Safe?

Vetiver is already so well known in so many countries that any serious threat from its use for erosion control would by now be obvious and widely reported. However, there are fertile types in India that could become hazardous if they are distributed.

A major reason for confidence in vetiver's safety is that the plant will seldom have to be introduced anywhere. It is already found throughout the tropics and has been there for at least a century. Apparently it has never spread in an uncontrolled manner or become a major nuisance. Possibly, such difficulties may be found as people investigate this plant more thoroughly, but, by and large, vetiver has not become a problem.

It is important, however, that only the right kind of vetiver be used. The types of South Indian origin apparently produce nonviable seeds and must be maintained by vegetative methods. Luckily, this is the fragrant-root type that has been spread throughout the tropics. In areas where it has been planted for decades, or in some cases for more than a century, it has seldom (if ever) spread from seed. In a few areas vetiver is reported as an escape, but even there it neither spreads rapidly nor is considered a nuisance.

On the other hand, types of vetiver grown from seeds introduced from northern India into the United States in 1989 *have* formed seeds and *have* germinated in areas adjacent to small plots in Georgia. This fertile type should not be introduced to new areas. It has long been used to protect canal banks in irrigated agriculture in northern India and the neighboring Terai of Nepal without becoming a pest or spreading uncontrollably from seed. Nonetheless, at this time, because of its potential hazard, only vegetative materials should ever be planted. Vetiver must never be propagated from seed.

Does Vetiver Reduce Runoff Water?

Yes, although by how much will depend on local conditions and is not now known.

A key claim for the vetiver system is that contour hedges of this grass can slow down and hold back moisture that would otherwise rush off and be lost to the slopes. This claim appears to be valid. For example, in Karnataka, India, farmers who plant vetiver can dig behind their hedges and find moist soil when their neighbor's land is parched. Similar observations have been made elsewhere. In many areas, this ability to hold moisture on the slopes, and thus increase infiltration, will likely boost crop yields and appeal to farmers, foresters, and civil authorities.

On the other hand, vetiver is not a barrier in the sense of being like a wall. Its hedges are more like tight filters laid across the land: they slow down runoff and retard its progress, but do not physically dam it up. Shallow runoff seeps through the lower part of the hedge; deeper runoff pours through (or even over) the upper parts. In this way, runoff is neither ponded nor concentrated but stays spread across the slope as it fell. The vetiver method contrasts with rock walls, dirt barriers, and other mechanical methods that hold, concentrate, or divert water off site.

Vetiver supporters see this ability to keep the water spread out across the slopes as a major advantage. They are probably right, at least for most locations. Much of today's most dramatic erosion is caused when runoff concentrates and scours out the land in "gully washers."

Will Farmers Adopt Vetiver?

In most places, yes—but not everywhere.

This simple, cheap, and easily understood way to control soil losses should be highly acceptable to farmers, foresters, and other potential users throughout the warm parts of the world.

For one thing, vetiver is not demanding. Its hedges are seldom more than 1 m wide and therefore take up little land. It generally requires relatively few plants to establish the hedges. It can be installed by farmers using their own labor and materials. It requires little or no maintenance once established, and the hedges can last for years.

For another, there are incentives for farmers to use vetiver. In most cases, the hedges will guide the bullock, the tractor, or the hoe along the contour. The resulting contour farming alone will likely conserve soil and improve moisture, and thereby give better yields, healthier plants, quicker growth, and more resistance to droughts and other hazards.

The appeal to self-interest could be powerful, obvious, and rapid; however, it is likely that certain farmers will resist the vetiver technique. Some will have farms so small that they cannot afford even the tiny strips occupied by vetiver hedges. Some will have narrow farms running up and down the slope (a common way of subdividing land among family members in some locations) so that hedges across the slopes would make plowing and other operations impractical or inconvenient. And some will care little for the land or its future (they may, for example, be tenant farmers, resentful of the landowner).

Will Farmers Dig Up Their Hedges?

Not often.

One criticism leveled against vetiver relates to the fact that its roots are worth money. Farmers, the critics charge, will dig up their hedges whenever cash becomes more important to them than erosion, which is to say almost always.

Reportedly, farmers have already dug up vetiver erosion hedges at least once in both Indonesia and Haiti. However, although such experiences will inevitably occur again, they will undoubtedly be few, far between, and seldom repeated. There are three major reasons for this:

- Difficult harvesting. Vetiver roots are threadlike strands that form lacy networks throughout a large volume of soil. To obtain the roots in quantity, therefore, vast amounts of dirt must be dug. Except in loose, light soils, the workload is enormous, especially considering the small amount of root obtained. Even in volcanic ash and sandy bottomlands, the task of digging vetiver roots is so difficult that it is often hard to find people willing to do it.

  Also, separating the soil from the intricate roots involves extreme drudgery. Commercial production is usually done only near waterways where the soil can be washed off.

- Lack of markets. The total world demand for vetiver oil is only about 250 metric tons per year (see sidebar, page 78). Moreover, there is little or no "elasticity" in the market, and the prospects for anyone to sell larger supplies are poor. Indeed, demand has been declining in recent years, and existing producers are more than capable of meeting any likely rebound. Currently, organized production is found in only three countries: Haiti, Indonesia (actually, only Java), and the tiny Indian Ocean island of Réunion. It is unlikely that they can be undercut because production costs in Indonesia and Haiti are probably as low as could be achieved anywhere. Further, the world's vetiver-oil buyers have contracts and relations with the traditional producing nations; they are therefore unlikely to turn to a new supplier without extraordinary financial or oil-quality inducements.

- Lack of facilities. To extract the oil from the roots is not easy or cheap—it demands a steam-distillation facility. Such a relatively expensive factory is unlikely to be an attractive investment in areas that don't already have one.

Does Vetiver Affect Adjacent Plants?

The answer is not clear. Depending on the crop and the level of stress at the site, some edge effects will undoubtedly be noted. However, neither vetiver nor its roots are inherently spreading, and in most cases, the effects should be minimal.

In theory, vetiver should compete with immediately adjacent plants for the use of water and nutrients, but in practice it doesn't seem to. Some yield reduction has been observed in the rows nearest the hedges in certain crops at certain sites (maize in India, for example). However, whether this results in an overall decrease in yield is as yet unclear—gains from the improved moisture levels across the whole field may well offset the losses at the upper and lower edges.

Certainly in some cases (cotton fields laced by vetiver hedges in South India, for example), no obvious edge effects are seen; the plants beside the hedges are as tall and productive as those elsewhere across the field. Perhaps this is because topsoil and organic debris washing off the fields accumulate behind the hedges and overcome any competition from the grass.

Is Vetiver Prone to Pests and Diseases?

No, but certain diseases and one potentially serious pest are known. In future these or other organisms might restrict the plant's usefulness to certain locations or climatic zones. Also, they might increase the need for the care and maintenance of the hedges.

Although vetiver plants are usually remarkably free of disease, several fungi have been identified in a few locations. For example, two smuts and a leaf blight (*Curvularia trifolii*) have been identified on vetiver plants in Bangalore, India. In Malaysia, three fungal diseases (tentatively identified as species of *Nigrospora*, *Curvularia*, and *Helminthosporium*) have been observed, but they do not appear to be virulent.

As for pests, a potentially nasty stem borer has been found in China. A severe but localized infestation of Eupladia grubs has been found in Africa. And termites are known to attack the dead stems, a particular problem in dry areas. None of these insects, however, seems to be a devastating threat to the plant's worldwide use.

Is Its Soil Adaptability as Broad as Claimed?

Vetiver's edaphic limits are unknown, and there are undoubtedly sites in which it will not grow. Nonetheless, the plant seems to tolerate a remarkable array of soil types.

There are well-documented examples where vetiver is growing in very adverse soils. For example:

- Coastal sand dunes in South Africa.
- Extremely acid soils (with pH as low as 4.0) in Louisiana.
- Highly alkaline soils (pH up to 11) at Lucknow, India.
- Black cotton clays (which heave and split and eject most plants) in Central India.
- Barren soils with little fertility or organic matter in South China and other places.
- Waterlogged soils in the black cotton clays, and even swamps, of India.
- Parched land. In the arid state of Rajasthan, India, vetiver is common.
- Saline soils in Australia.

Although vetiver can survive on many hostile sites, it should not be expected to grow there with its full vigor. On many poor soils the plant will probably be difficult to establish, its growth will be retarded, and its ability to form a hedge to catch erosion will be long delayed. Some fertilization will often be required on such sites to create and maintain fully functioning hedges.

What Range of Climates Can the Plant Withstand?

Details are unknown, but vetiver's climatic limits seem to be remarkably broad. The plant, however, cannot reliably withstand freezing conditions.

Heat seems to be no barrier. Vetiver is reported to grow vigorously under some of the most torrid conditions faced by farmers anywhere.

On the other hand, vetiver is definitely limited by cold. It can take a mild freeze, but even this may be misleading. An erosion-control barrier must be able to function for years to be reliable; should a hedge die with even infrequent cold snaps, all the soil accumulated behind it would be left vulnerable.

From experiences in New Zealand and southern England, it appears that vetiver will succeed only where the sun is hot and bright. Maritime temperate locations may prove too cool and cloudy for the hedges to form up quickly.

Aren't There Other Species that Can Do the Same Job?

So far, no other plant with all of vetiver's major qualities has been found. However, some have certain of its qualities.

It would be most desirable to have additional species to use in narrow vegetative barriers. They could extend the geographic range of hedgerow erosion control, give farmers more options, and provide substitutes for vetiver should it ever get struck down by epidemic disease.

Effectiveness of other species is also a management question. Napier grass might be as efficient as vetiver if kept as a much wider hedge, but it would require significantly more care, expense, and labor—as well as replanting every few years.

Isn't It a Threat to Have Erosion Control Based on a Monoculture?

Yes. At present, however, there seems to be no alternative for use in narrow hedgerow barriers against erosion.

The job vetiver can do is so important that it would be absurd not to take advantage of it just because some unknown disease may possibly break out in future. Research should be maintained on alternatives (see Appendix B), but vetiver is currently the only practical species for low-maintenance, single-plant hedgerow erosion barriers.

Although the vetiver found almost everywhere in the tropics is the same species, it has been scattered in myriad locations for a long time, and it is possible that the germplasm is not as genetically uniform as might appear.

Will the Plant Foster Diseases or Pests that Might Attack Crops?

Perhaps, but there is no evidence for it at present.

Vetiver belongs to the same subtribe of the grass family as maize, sorghum, lemongrass, and citronella. It is therefore not unexpected that it will be susceptible to some of the same maladies. A more serious fear, however, is that bands of this perennial across the fields might act as sources of infections that damage the crops.

To date, no one has reported that vetiver hedges can act this way. They of course take up only a small area and are unlikely to produce massive populations of pests even in the worst of cases. Thus, although they may provide cozy places for pests to overwinter, the probability of serious problems seems slight. It is likely, also, that management practices (such as periodically burning and topping the hedges) will further minimize any threats.

Is Vetiver Sterile?

Certain types appear to be sterile in that they do not spread under normal circumstances. However, whether they produce truly sterile seed or whether they germinate only under special and unusual conditions is not clear.

It is legitimate to fear that any grass belonging to vetiver's subtribe could become a weed. Others have become so: shattercane and Johnson grass are just two examples. But the South India form of vetiver is not considered a weed in the dozens of places where it has been planted around the world. Also, the weedy members of the subtribe tend to be annuals. The perennials, such as vetiver, show a marked tendency toward low seed-set and an absence of female flowers. This makes them less aggressive.

It is important, however, that any vetiver planted be of the South India type. This is the type now spread widely through the tropics. As previously noted, a second type growing wild in North India is flowering and fertile and produces seedlings under conditions of high humidity and high soil moisture.

Although in one or two cases even the "sterile" vetiver has been declared a weed, it was a mild nuisance rather than a spreading terror. In hundreds of locations, the plant is considered benign at worst and highly beneficial at best. Indian farmers, for example, do not see it as a weed, even where it has been on their farms for centuries.

Generally, vetiver grows only where people plant it.

How Long Will a Hedge Last?

Vetiver is a perennial and, at least in some locations, hedges formed from it seem capable of surviving for decades.

This grass is remarkably persistent. On the island of St. Vincent in the Caribbean, vetiver barriers have been used for 50 years and some are still active even without maintenance. On an experiment station in northern Zambia, vetiver is reportedly still in the same lines that were planted 60 years ago. Some boundary strips in vetiver's native region of India are purported to be at least 200 years old.

Do Vetiver Hedges Require Maintenance?

There are mixed reports on this, but probably a rough trimming every year or two will be needed to keep most hedges in the best operating condition.

Although vetiver stays in place without requiring attention, it tends to break up into clumps if not trimmed. This usually takes a long time, however. Fiji's sugarcane fields offer an example where hedges given little or no maintenance produced excellent results for more than 30 years.

On the other hand, it is reported in St. Lucia that the hedges get rough and ragged unless they are periodically trimmed. One hedge in St. Lucia showed some gullying where gaps had formed, but it had been abandoned for probably more than 30 years. In addition, the flat surface of the terrace showed signs of sheet erosion (rocks on top of the soil) because water had drained too fast through the thinned hedge.

Trimming the hedges is usually not difficult. Generally, farmers merely run a plow along the edge to cut off any spreading tillers and cut back the tops with a machete. The effort can provide at least small amounts of forage, mulch, animal bedding, or thatch.

Is It Expensive?

The vetiver method is inherently low cost.

Compared to terracing, bunding, or land leveling, vetiver is very inexpensive. In the red soils area of China, for example, the mechanical techniques cost, on average, $900 per hectare; vetiver, by contrast, costs less than $200. A similar ratio has been measured in India, where the cost for bunds was $60 per hectare and for vetiver less than $20 per hectare.

These figures were for initiating a vetiver program. They include costs for nurseries, the production of planting materials, the planting process itself, and perhaps some early maintenance (watering, fertilizing) to give a hedge a head start. As the critical mass of hedge builds up, the farmers obtain planting material from their own plots at no cost.

Once established, the hedge requires little or no expense. Constructed soil-conservation measures, on the other hand, require constant, and sometimes extremely costly, maintenance.

When the whole system is taken into account, vetiver in many places may more than pay for itself in improved crop yields. Yield increases of 25–60 percent have been recorded for contour cultivation of cotton, sorghum, and other crops in India.

Is It Easy to Establish?

In most places in the tropics it should be easy to establish vetiver hedges. In some, however, careful tending and perhaps years of delay will be required before the hedges are fully serviceable.

Vetiver is easily planted by hand, but machines (tobacco- or vegetable- or tree-seedling planters, perhaps) could also be used. The main work is to break up sprigs and to set them in the soil.

Any shortage of planting materials will, in most cases, be temporary. Over the years, a little patch of nursery can produce planting materials that can cover vast areas. Under irrigation and fertilization, a single hectare of nursery can, within a few months, provide enough slips to plant about 150 km of hedgerows, which, depending on slope, could protect up to 450 hectares.

Although vetiver establishes amazingly well where conditions are at least moderately favorable, delays and disappointing results can occur where conditions are difficult. For example, it requires good management to get vetiver established in very dry areas. Here, several years may pass before the plants have grown together to form tight hedges.

Must the Hedges Be Unbroken?

No.

Whereas a gap in an earth bund is catastrophic because the impounded body of water pours through, a vetiver hedge is not a solid barrier and it need not be perfect to work. Because the hedge filters water and doesn't dam it or divert it, it is tolerant of an odd gap. Nonetheless, a continuous hedge is desirable, and there are some reports of "gappy" hedges exacerbating erosion by channelling the water.

It is tolerant of other imperfections as well. For example, the hedge need not be exactly on the contour nor have perfectly even growth. Any low points will tend to gather more soil and trash and therefore will respond with more sediment accumulation, eventually forming a level terrace. Bunds and berms, on the other hand, must be on the exact contour because they function by damming or diverting the runoff water.

Wouldn't a Plant with More Uses be Better?

Maybe, in some situations.

Some have said that to be acceptable to farmers, any erosion-blocking plant should be more palatable to animals. That way, they say, people will put in the hedges to feed their livestock. Without some such incentive, they claim, farmers won't respond.

We are not convinced of this argument. Indeed, if a grass barrier is too palatable to livestock, it may lose its effectiveness. Animals are found almost everywhere in Third World countries, and wildlife or livestock will reduce an edible erosion barrier to ground level, especially in the dry season when feeds are in short supply. The advantage of vetiver, however, is that once it is established, livestock do not affect its usefulness.

Actually, though, vetiver yields several products that can be sold without damaging its effectiveness; for example, its stems and leaves can be used for making ropes, hats, brushes, thatch, mats, and fuel. They also make excellent mulch and animal bedding. In some places, vetiver is an important medicinal plant and, although the whole plants are shunned by animals, young vetiver leaves are palatable to livestock and can be used as feed.

Does Vetiver Cause Erosion?

No.

It may seem ironic, but it is nonetheless true that vetiver cultivation is barred from certain parts of Indonesia solely because it "causes soil erosion." This had led several people to denounce all efforts to promote the wider use of vetiver. After all, they point out, more vetiver is grown for oil in Indonesia than anywhere else except Haiti; surely Indonesians must know the plant and its performance.

Closer inspection, however, shows that the culprit is not vetiver itself but the specific way farmers grow and harvest it there. It is so difficult to dig up the roots that the plants are grown in special sites where the soil is extremely light. Only there can the farmers obtain the roots with a reasonable effort. When the time comes, they rip the plants out, leaving behind trenches of loose dirt that could hardly be more erodible if designed for the purpose. Some farmers even place their rows up and down (rather than across) the slopes. This allows the rain to scour the land in an even more disastrous manner.

When vetiver hedges are established on the contour and left in place, there is no evidence that they cause erosion.

Is It Difficult to Get Rid of?

No.

Vetiver is easily killed by slicing off the crown with a shovel or other implement. It is also easily eradicated by systemic herbicides. It is so easy to remove that in some places it is used in crop rotations as a fallow crop.

Why Hasn't Vetiver Been Widely Used Before?

We don't know why.

Over the past 50 years, the main approach to worldwide erosion control has been oriented toward engineered systems, such as terraces, bunds, or contour drains. The use of vegetative systems is generally not attractive to conservation specialists trained in engineering techniques.

Another major approach worldwide is the use of cultural systems such as crop rotations, contour planting, ridge planting, and mulches. Erosion-control hedge technology has been lying dormant.

5
The Plant

Vetiver belongs to the same part of the grass family as maize, sorghum, sugarcane, and lemongrass.[1] Its botanic name, *Vetiveria zizanioides* (Linn) Nash, has had a checkered history—at least 11 other names in 4 different genera have been employed in the past. The generic name comes from "vetiver," a Tamil word meaning "root that is dug up." The specific name *zizanioides* (sometimes misspelled *zizanoides*) was given first by the great Swedish taxonomist Carolus Linnaeus in 1771. It means "by the riverside," and reflects the fact that the plant is commonly found along waterways in India.

NATURAL HABITAT

For a plant that grows so well on hillsides, vetiver's natural habitat may seem strange. It grows wild in low, damp sites such as swamps and bogs.

The exact location of its origin is not precisely known. Most botanists conclude that it is native to northern India; some say that it is native around Bombay. However, for all practical purposes, the wild plant inhabits the tropical and subtropical plains throughout northern India, Bangladesh, and Burma.

THE TWO TYPES

It is important to realize that vetiver comes in two types—a crucial point because only one of them is suitable for use around the world. If the wrong one is planted, it may spread and produce problems for farmers.

[1] The actual family tree: Family—Graminae (Poaceae), Subfamily—Panicoideae (Andropogonidae), Tribe—Andropogoneae, Subtribe—Sorghinae.

The two are:

- A wild type from North India. This is the original undomesticated species. It flowers regularly, sets fertile seed, and is known as a "colonizer." Its rooting tends to be shallow, especially in the damp ground it seems to prefer. If loosed on the world, it might become a weed.
- A "domesticated" type from South India. This is the vetiver that has existed under cultivation for centuries and is widely distributed throughout the tropics. It is probably a man-made selection from the wild type. It is nonflowering, nonseeding (or at least nonspreading), and must be replicated by vegetative propagation. It is the only safe type to use for erosion control.

It is not easy to differentiate between the two types—especially when their flowers cannot be seen. Over the years, Indian scientists have tried to find distinguishing features. These have included differences in:

- Stems. The South Indian type is said to have a thicker stem.
- Roots. The South India type is said to have roots with less branching.
- Leaves. The South India type apparently possesses wider leaves (1.1 cm vs 0.7 cm, on average).[2]
- Oil content. The South India type has a higher oil content and a higher yield of roots.
- Physical properties. Oil from the wild roots of North India is said to be highly levorotatory (rotates the plane of polarized light to the left), whereas that from the cultivated roots from South India is dextrorotatory (rotates polarized light to the right).
- Scent. The oils from the two types differ in aroma and volatile ingredients.

Whether these differences are truly diagnostic for the two genotypes is as yet unclear. However, at least one group of researchers consider that the two vetivers represent distinct races or even distinct species.[3] Perhaps a test based on a DNA profile will soon settle the issue.

PHYSIOLOGY

Like its relatives maize, sorghum, and sugarcane, vetiver is among the group of plants that use a specialized photosynthesis. Plants

---

[2] Sobti and Rao, 1977.
[3] CSIR, 1976.

employing this so-called C_4 pathway use carbon dioxide more efficiently than those with the normal (C_3 or Calvin cycle) photosynthesis. For one thing, most C_4 plants convert carbon dioxide to sugars using less water, which helps them thrive under dry conditions. For another, they continue growing and "fixing" carbon dioxide at high rates even with their stomata partially closed. Since stomata close when a plant is stressed (by drought or salinity, for instance), C_4 plants tend to perform better than most plants under adversity.

The vetiver plant is insensitive to photoperiod and grows and flowers year-round where temperatures permit. It is best suited to open sunlight and will not establish easily under shady conditions. However, once established, plants can survive in deep shade for decades. They tolerate the near darkness under rubber trees and tropical forests, for example.

ARCHITECTURE

In its general aspect a vetiver plant[4] looks like a big, coarse clump grass—not very different from pampas grass, citronella grass, or lemongrass. It can, however, grow to be very tall. Under favorable conditions the erect stems (culms) can reach heights of 3 m.

For purposes of erosion control, vetiver has a number of singular architectural and anatomical features:

- Habit. The plant has an erect habit and keeps its leaves up off the ground. This seems to be important in allowing the hedge to close up tight, and it also allows crops to be grown next to the plant.
- Resistance to toppling. Unlike many grasses, vetiver is "bottom heavy." It shows no tendency to fall over (lodge), despite its very tall culms.
- Strength. The woody and interfolded structures of the stems and leaf bases are extremely strong.
- Year-round performance. Although vetiver goes dormant during winter months or dry seasons, its stems and leaves stay stiff and firmly attached to the crown. This means that the plant continues stopping soil, even in the off-seasons or (at least for some months) after death.
- Self-rising ability. As silt builds up behind a vetiver plant, the crown rises to match the new level of the soil surface. The hedge is thus a living barrier that cannot be smothered by a slow rise of sediment. Like dune grasses at the beach, it puts out new roots as dirt builds up around its stems.
- "Underground networking." Vetiver is a sod-forming grass. Its clumps grow out, and when they intersect with neighboring ones they

[4] From here on, and everywhere else in this report, we are referring exclusively to the South India (domesticated) type unless specifically noted.

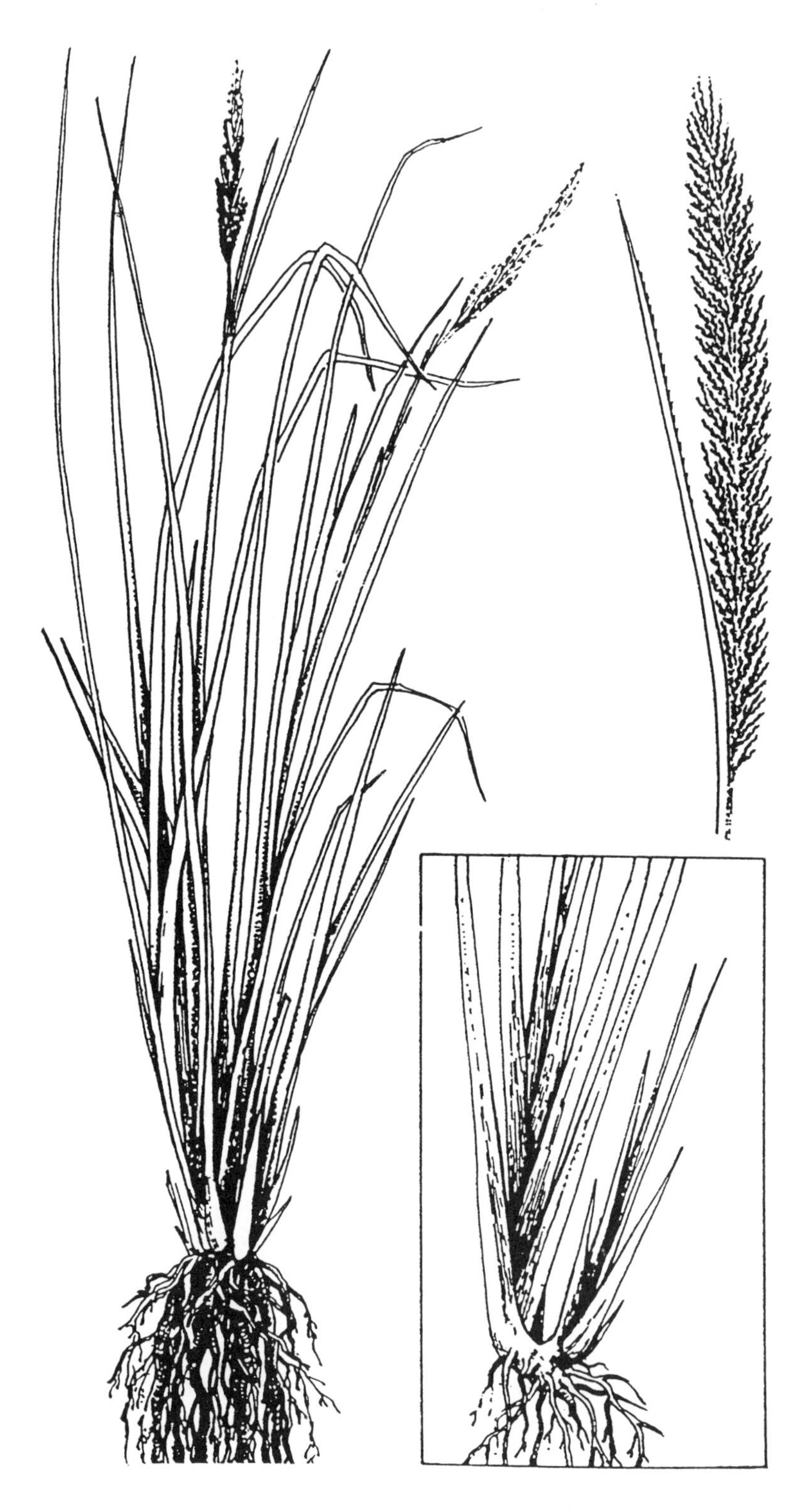

intertwine and form a sod. It is this that makes the hedges so tight and compact that they can block the movement of soil.

- Clump integrity. For all practical purposes, vetiver has no running rhizomes or stolons.[5] This, too, helps keep the hedge dense and tight. The clumps do not readily die out in the center. Unlike most other clump grasses, even old vetiver plants seldom have empty middles.[6]

Crown

The crown of the plant is generally a few centimeters below the surface of the ground. It is a "dome" of dead material, debris, and growing tissue, much of it a tangled knot of rhizomes. These rhizomes are very short—1 cm or less—and are often turned back on themselves. It is apparently for this reason that vetiver stays in clumps and does not spread across the land.

To separate the slips for planting, the often massive crown is cut apart. It is sometimes so huge that it has to be pulled out of the ground with a tractor and cut up with axes. In nurseries, however, young slips are easily separated.

Leaves and Stems

It is, of course, the leaves and stems that are crucial in this living-hedge form of erosion control. Vetiver leaves are somewhat like those of sugarcane, but narrower. Although the blades are soft at the top, the lower portions are firm and hard.

On some vetiver types the leaves have edges sharp enough to cut a person. Actually, this is due to tiny barbs. There is a lot of variability, however: some plants are fiercely barbed, some not. The ones used for oil and erosion control tend to be smooth edged. Topping the plants is an easy way (at least temporarily) to remove the bother of the barbs.

The leaves apparently have fewer stomata than one would expect, which perhaps helps account for the plant withstanding drought so well.

It is the stems that provide the "backbone" of the erosion-control barrier. Strong, hard, and lignified (as in bamboo), they act like a wooden palisade across the hill slope. The strongest are those that bear the inflorescence. These stiff and canelike culms have prominent nodes that can form roots, which is one of the ways the plant uses to rise when it gets buried. (It also moves up by growing from rhizomes on the crown.)

Throughout their length, the culms are usually sheathed with a

---

[5] Actually, it does have small rhizomes, which, because they are folded back on themselves, don't run outwards in the normal manner.

[6] This is especially the case if they have been periodically topped.

Even experienced plant scientists are surprised by the shape and massiveness of vetiver's roots. The thin and lacy roots grow downwards rather than sideways, and form something like a curtain hanging in the soil. When two plants are side by side, their roots interlock into an underground network. This combination of features anchors a hedge so firmly that even the strongest floods can seldom undermine it or wash it out. Moreover, the roots fall away at a steep angle, and this conical form perhaps explains why vetiver appears not to affect nearby crops. (P.K. Yoon)

leaflike husk. This possibly shields them from stresses—salinity, desiccation, herbicides, or pestilence, for example.

Flowers

The flower (inflorescence) and seedhead are very large: up to 1.5 m long. Both are brown or purple in color. The flower's male and female parts are separated. As in maize, florets in the upper section are male and produce pollen; in vetiver, however, those below are hermaphrodite (both male and female).

Roots

Perhaps most basic to this plant's erosion-fighting ability is its huge spongy mass of roots. These are not only numerous, strong, and fibrous, they tap into soil moisture far below the reach of most crops.

They have been measured at depths below 3 m and can keep the plant alive long after most surrounding vegetation has succumbed to drought.

The massive, deep "ground anchor" also means that even heavy downpours cannot undermine the plant or wash it out. Moreover, because the roots angle steeply downwards, farmers can plow and grow their crops close to the line of grass, so that little cropland is lost when the hedges are in place.

The roots can grow extremely fast. Slips planted in Malaysia produced roots 60 cm deep in just 3 weeks.[7]

FERTILITY

One of vetiver's great benefits, of course, is that once it is planted it stays in place. It is therefore not pestiferous and seldom spreads into neighboring land.

Actually, though, seeds are often seen on the plant. Why they fail to produce lots of seedlings is not known. Perhaps they are sterile.[8] Perhaps they are fertile but the conditions for germination are seldom present. Or perhaps people just haven't looked hard enough.

Under certain conditions some seeds are indeed fertile. These conditions seem to be most commonly found in tropical swamps. There, in the heat and damp, little vetivers spring up vigorously all around the mother plant.

ECOLOGY

Vetiver is an "ecological-climax" species. It outlasts its neighbors and seems to survive for decades while (at least under normal conditions) showing little or no aggressiveness or colonizing ability.

VETIVER OIL

The oil emits a sweet and pleasant odor. It is used particularly in heavy oriental fragrances. Although primarily employed as a scent, it is so slow to evaporate from the skin that it is also used as a fixative that keeps more volatile oils from evaporating too fast. Because it does not decompose in alkaline medium, vetiver oil is especially good for

[7] Yoon, 1991.

[8] Researchers who examined spikelet and pollen fertility in 75 clones collected from widely different geographical locations within India found that 5 clones failed to "flower." Of the remaining 70, female (pistillate) sterility ranged from 30 percent to 100 percent; male (pollen) sterility ranged from 2 percent to 100 percent. Some, notably those of South Indian origin, could be maintained only by vegetative methods because they produced no seeds under natural pollination or under hand pollination, despite high pollen fertility. (Ramanujam and Kumar, 1963b.) The measurements were made under New Delhi conditions.

Vetiver Oil

The oil in vetiver roots has a pleasant aroma. The perfumery industry describes it as "heavy," "woody," or "earthy" in character. It is obtained by steam distilling the roots and is used in fine fragrances and in soaps, lotions, deodorants, and other cosmetics. Occasionally its scent dominates a perfume, but more often it provides the foundation on which other scents are superimposed.

Haiti, Indonesia (actually only Java), and Réunion (a French island colony in the Indian Ocean) produce most of the world's vetiver oil. China, Brazil, and occasionally other nations produce smaller quantities. Réunion produces the best oil, but Haiti and Indonesia produce the most. Haitian oil is distinctly better than Indonesian and not far behind the Réunion oil (known in the trade as "Bourbon vetiver") in quality.

Although reliable statistics are unavailable, world production of vetiver oil is currently about 250 tons a year. Annual consumption is estimated to be:

United States	100 tons
Francc	50 tons
Switzerland	30 tons
United Kingdom	20–25 tons
Japan	10 tons
Germany	6 tons
Netherlands	5 tons
Other	30–40 tons

It seems unlikely that demand will increase beyond these figures, even to match population growth. In recent decades, the international perfumery industry has generally decreased its use in new products. This decision was taken primarily because Haiti manipulates the price of its oil, Indonesia's oil is indifferent and variable in quality, and Bourbon oil is expensive. A completely synthetic vetiver oil cannot be manufactured at a realistic price, but alternative materials such as cedarwood oil can be substituted. For these reasons, world vetiver-oil consumption is likely to remain roughly at current levels.

Also, some countries have given up producing vetiver oil. For example, Guatemala, which was once an exporter, no longer produces any (even for local use), and Angola (until the early 1970s a regular supplier to the international market) has given up as well.

Adapted from S.R.J. Robbins, 1982

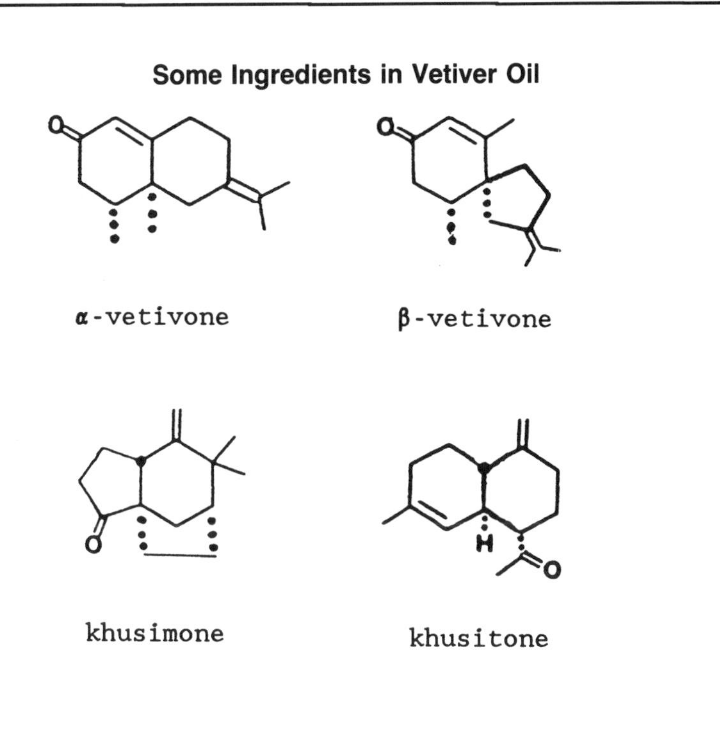

American Vetiver Oil

The United States is not known as a vetiver-oil producer, but as this book was about to be printed we learned that Texas farmers Gueric and Victor Boucard are perhaps the most advanced vetiver growers of all. Since 1972, the Boucards and their late father have been designing implements to plant, harvest, and process vetiver for its oil. These days the Boucards grow the crop on as much as 40 hectares (depending on market prices). All the necessary steps—from large-scale nursery operations to digging to roots—are mechanized. (A modified rock-picker has proved ideal for tearing up the roots.)

To us, the surprising thing is that the grass survives in this far-from-tropical location (nearly 30°N latitude and above 500 m). However, Gueric Boucard indicates that this is no problem: "I just cut them off near the ground each fall," he says, "and they have survived temperatures down to −7°C winter after winter, and one cold snap below −12°C for several hours."

scenting soaps. In at least one country it also serves as a flavoring, primarily to embellish canned asparagus or sherbets.

As already noted, the dried roots are used in India to prepare the traditional "khus-khus" screens. When moistened, these both cool and scent the air passing through, and they are believed to protect people against insect pests as well.

The oil occurs primarily in the roots, but traces of it in the foliage may nonetheless account for the plant's inherent resistance to pests and diseases. The oil is known to repel insects, for example. People in India and elsewhere have long placed vetiver root among their clothes to keep insects away. There seems to be validity in this. In experiments, vetiver root has protected clothes from moths, heads from lice, and bedding from bedbugs. The oil repels flies and cockroaches as well and may make a useful ingredient in insect repellents.

The oil is extremely complex, containing more than 60 compounds. In the main, these are bicyclic and tricyclic sesquiterpenes—hydrocarbons, alcohols, and carboxylic acids. Those that repel insects are minor constituents, including α-vetivone, ß-vetivone, khusimone, and khusitone (see illustration, previous page).

ENVIRONMENTAL LIMITS

As noted previously, the plant's environmental limits are unknown. They are, however, surprisingly broad.

As far as moisture is concerned, an established vetiver plant can grow in sites where annual rainfall is perhaps as little as 200 mm. At the other extreme, it also shows tremendous growth where annual rainfall is 3,000 mm. And in Sri Lanka it grows where rainfall is as much as 5,000 mm.

As far as temperature is concerned, this tropical species can take any amount of heat but cannot be counted on to survive subfreezing conditions. For example, plants in Georgia (USA) survived when soil temperatures reached −10°C without apparent damage but died when soil temperatures reached −15°C.

DISEASES

Vetiver is remarkably free of disease.[9] However, *Fusarium* (the most widespread cause of rotting in fruits and vegetables) reportedly attacks it, notably during rains.[10]

[9] In a letter to us, plant pathologist P.S. Teng, of the International Rice Research Institute in the Philippines, wrote: "I must have examined some 50 vetiver clumps and was 'disappointed' to find no disease or insect infestations!"

[10] Drenching the soil with fungicide (one percent Bordeaux mixture, for example) is said to reduce the incidence of this serious fungal disease. CSIR, 1976.

Perhaps of greater import is the leaf blight caused by *Curvularia trifolii*. This disease of clover and other crops may attack vetiver also during the rainy season. Malaysian researchers recommend that growers top the plants (at a height of 20–30 cm) to remove any infected foliage. Copper-based fungicides such as Bordeaux mixture also control this blight.

In Malaysia, a detailed investigation of vetiver has located yet more fungal species.[11] These had little effect on the plant itself, but they might eventually prove troublesome in crops grown near vetiver hedges. They include the following species:

- *Curvularia lunata* (causes leaf spot in oil palm)
- *C. maculans* (causes leaf spot in oil palm)
- *Helminthosporium halodes* (causes leaf spot in oil palm)
- *H. incurvatum* (causes leaf spot in coconut)
- *H. maydis* (causes leaf blight in maize)
- *H. rostratum* (causes leaf disease in oil palm)
- *H. sacchari* (causes eye spot in sugarcane)
- *H. stenospilum* (causes brown stripe in sugarcane)
- *H. turcicum* (causes leaf blight in maize).

PESTS

Termites sometimes attack vetiver, but seemingly only in arid regions. Except where the termite mound covers the whole plant, only dead stems in the center of stressed plants are affected. Normally no treatments are required.

In at least one location in India, grubs of a beetle (*Phyllophaga serrata*) have been found infesting vetiver roots.[12]

Perhaps the most serious pest threat comes from stem borers (*Chilo* spp.). These were found in vetiver hedges in Jianxi Province, China, in 1989. In Asia and Africa, some of these moth larvae are severe cereal pests (for example, the rice borer of Southeast Asia and the sorghum borer of Africa).

Until this potential problem is better understood, vetiver plantings should be carefully monitored in areas where stem borers are a problem. This is both to protect the hedges and to prevent them from providing safe havens for these crop pests. A severe pruning seems to keep the larva from "overwintering" in the vetiver stems and a timely fire might also be beneficial.[13]

---

[11] Yoon, 1991.

[12] CSIR, 1976.

[13] Because vetiver is unfazed by fire, this technique might be developed into a method for trapping and destroying these pests.

Root-Knot Nematodes

Vetiver has outstanding resistance to root-knot nematodes. In trials in Brazil it proved "immune" to *Meloidogyne incognita* race 1 and *Meloidogyne javanica*.[14]

PROPAGATION

Currently, vetiver is propagated mainly by root division or slips. These are usually ripped off the main clump and jabbed into the ground like seedlings. Although the growth may be tardy initially, the plants develop quickly once roots are established. Growth of 5 cm per day for more than 60 days has been measured in Malaysia. Even where such rapid growth is not possible, the plants often reach 2 m in height after just a few months.

It is easy to build up large numbers of vetiver slips. The plant responds to fertilizer and irrigation with massive tillering, and each tiller can be broken off and planted. It is important to put the nurseries on light soil so the plants can be pulled up easily.

Planting slips is not the only way to propagate vetiver. Other vegetative methods follow:

- Tissue culture. Micropropagation of vetiver began in the late 1980s.[15]
- Ratooning. Like its relative sugarcane, the plant can be cut to the ground and left to resprout.
- Lateral budding. Researchers in South Africa are having success growing vetiver "eyes" (intercalary buds on surface of crown) in seedling dishes.[16]
- Culms. As noted in chapter 2, young stems easily form new roots. This can be an effective means for propagating the plant. Laying the culms on moist sand and keeping them under mist results in the rapid formation of shoots at each node.[17] This is an effective way to propagate new plants from hedge trimmings.
- Cuttings. One Chinese farmer has successfully grown vetiver from stem cuttings. The cuttings, each with two nodes, are planted at a 60° angle and then treated with a rooting hormone—in this case, IAA

[14] Marinho de Moura et al., 1990.

[15] A British company, MASDAR, sells tissue-cultured vetiver plantlets based on a clone from Mauritius. It has shipped them to Sudan, Zambia, Somalia, Nigeria, New Zealand, and other nations.

[16] Information from A. Tantum.

[17] Yoon, 1991.

(indole acetic acid). He achieved 70 percent survival. An interesting point was that the original stems were cut in December, buried in the ground over the winter, and the cuttings were made in early spring and planted in April.

Hedge Formation

Normally, hedges are established by jabbing slips into holes or furrows. They can be planted with bullock, trowel, or dibbling stick. In principle, at least, the techniques and machines developed for planting tree or vegetable seedlings could also be employed.

To establish the hedge quickly, large clumps can be planted close together (10 cm). On the other hand, when planting material is scarce, slips can be spaced as far apart as 20 cm. In this case, the hedge will take longer to close.

Prolonged moisture is highly beneficial for the quick establishment of the hedge. For best results, fresh and well-rooted slips, preferably containing a young stem, should be planted early in the wet season (after the point when there is a good chance the rains will continue). In drier areas it is helpful to plant them in shallow ditches that collect runoff water. For the most rapid establishment of vetiver lines, weeding should be done regularly until the young plants take over. Clipping the young plants back stimulates early tillering and makes the hedge close up faster.

Management

Usually, little management is needed once the hedge is established. However, cutting the tops of the plants produces more tillering and therefore a denser hedge.

CONTROLLING VETIVER

Vetiver is a survivor. It is difficult to kill by fire, grazing, drought, or other natural force. However, if necessary, it can be eliminated by slicing off the crown. Because the crown is close to the surface, it can be cut off fairly easily with a shovel or tractor blade. Also, although the plant is resistant to most herbicides, it succumbs to those based on glyphosate.

6

Next Steps

In light of the massive environmental destruction now caused by erosion, any system that retards soil loss would seem to be a candidate for instant and widespread use. Indeed, if the power of the vetiver system is as great as now appears, in a few decades the world could see thousands of kilometers of vetiver hedges in a hundred different countries, in climates from lush to harsh, and in sites from verdant to sparse.

But the experiences so far are limited, and many uncertainties remain. It is time, therefore, for a wide-ranging exploration of this new, and seemingly revolutionary, technique.

In this chapter we identify some of the important, as well as some of the interesting and challenging, actions that could be taken to help vetiver progress in an orderly, responsible, and yet rapid manner.

CONTINUATION OF WORLD BANK EFFORTS

As discussed earlier, it has been World Bank agriculturists who have reintroduced the vetiver method of erosion control. Their enthusiasm and energy have stimulated people all over the world and it will be of inestimable value to the testing and adoption of vetiver if these activities continue.

The World Bank staff also created the Vetiver Network—a service that collects and disseminates vetiver information and maintains an address list of more than 2,000 vetiver specialists and aficionados. With interest in this grass rising rapidly, continued support for this activity is essential. Such a service prevents duplication of effort and ensures that interested people receive constant updates of information on experiences with vetiver in other parts of the world.

The fostering of a new technology like this is not, however, a normal World Bank operation, and it would be prudent for other organizations to initiate a complementary operation to take over should the abandonment of this project occur.[1]

PRACTICAL STUDIES

It is clear that vetiver could become a vital component of land use throughout the warmer parts of the world. It might be a low-cost way to protect billions of dollars of investment already made in agriculture and forestry, not to mention roads, dams, and other public works throughout Africa, Asia, and Latin America. It also could become a backstop built into future projects as a way to help protect the environment from many soil disturbances.

At present, however, that is all speculation. It is critical for countries to establish vetiver trials quickly, which would serve to show local decision makers what this grass has to offer their programs and projects.

To expedite, motivate, and assure success in such a massive number of trials, we suggest consideration of what could be called a "vetiver SWAT team." This might involve a small number of vetiver specialists, brought in on short-term assignment to show local authorities how and where to put in small vetiver demonstration trials.

These trials might be incorporated into specific projects dealing with topics such as those discussed below.

Agriculture

In farming areas, trials could be particularly effective. With vetiver hedges reducing rainfall runoff, farmers should encounter more moisture in their soils so that crops produce higher yields and tolerate drought better. Given a successful demonstration, the word is likely to spread rapidly from farmer to farmer. Thus, not only the environmental stability of an area but also its productivity may increase.

Forestry

Foresters are likely to be impressed as well. By holding soil and moisture on site and by providing windbreaks, vetiver strips would be

[1] The Nitrogen-Fixing Tree Association is perhaps a good model. One professor, some part-time help from graduate students, and support from eager volunteers in a dozen nations was all that was needed to get leucaena and other nitrogen-fixing trees established as prime reforestation candidates throughout the tropics. Another excellent model is the International Ferrocement Information Center, located at the Asian Institute of Technology near Bangkok, Thailand. It provides similar services, with verve and style, to people interested in the construction material called "ferrocement." The key is the enthusiasm, commitment, and drive of its staff, not its size or budget.

particularly valuable in the early stages of tree growth. At present, a huge proportion of tree-planting projects are failing because of dismal rates of establishment and survival. Vetiver might also act as a barrier against ground fires and creeping grasses, both of which often devastate young tree plantings.

Public Works

At least initially, engineers are likely to be apathetic (if not apoplectic) toward the idea of using a grass for erosion control. The past 50 years or so have seen the rise in popularity of erosion controls based on bulldozers, land surveys, and engineered systems such as terraces, berms, bunds, and contour drains. By contrast, a strip of grass seems puny and insignificant.

However, demonstrations are likely to persuade everyone that vetiver hedges can protect and enhance the performance, as well as extend the useful lifetimes, of many structures made of steel, concrete, or asphalt. For instance, vetiver could help protect footpaths, railroads, and road cuts from washouts and slips. In addition, it has potential benefits for wastewater treatment and flood-control facilities. Further, because it can withstand lengthy submergence (more than 2 months has been reported), it can be planted along the edges of dikes, irrigation canals, bridges, and dams to prevent scouring.

Siltation

Most observers have despaired of the possibility of radically reducing the vast amounts of silt washing every day into ditches, canals, reservoirs, rivers, harbors, estuaries, and other waterways worldwide. This would be a massive, perhaps impossible, task for any erosion-control technique. Vetiver, though, just might work. Rows of this grass across critical watersheds should reduce silt buildup downstream. Indeed, although huge plantings would be required, government authorities might find the expense and effort far outweighed by the financial benefits of extending the useful life of multimillion-dollar water projects, not to mention the protection of wetlands, coral reefs, and other vital economic environments.

The advantage of the vetiver system is the fact that it is applicable on a wide scale with little equipment, planning, or logistics. Further, it is likely to appeal to those actually occupying the land. This is vital: farmers and foresters may well protect their land with little or no urging—not for soil conservation per se, but for the increased yields fostered by the moisture held back by the vetiver hedges, not to mention the by-products of vetiver—fire control, forage, mulch, thatch, and so on.

Flood Control

Wherever in the tropics localized flash flooding is a problem, vetiver could be part of the solution—especially when the flooding is caused by denuded watersheds that can no longer soak up and hold back the rainfall and runoff.

Desertification

Although rain falls infrequently in arid lands, it often adds up to considerable amounts. Moreover, desert rains are often intense deluges and the water rushes away uselessly down wadis and washes. Paradoxically, the water the desert needs so desperately is lost by flooding.

Vetiver barriers across those wadis and washes would likely capture the floodwater to recharge the thirsty aquifers beneath. Walls of vetiver would also hold back silt in which crops might grow vigorously.

In addition, vetiver hedges might prove excellent as windbreaks in desertifying areas. Already, palm fronds are used for wind protection in the Sahel, for example (see Appendix A). These palm-frond fences look somewhat like rows of dead vetiver, but it is likely that the living, growing, real thing would be far better—especially given vetiver's ability to withstand undermining and to "rise up" as sand or soil collects around it.

Sustainable Agriculture

There is currently a great interest in keeping agriculture productive and self-sustaining. Although vetiver has seldom been considered in the many parleys, papers, projects, and predictions, keeping the soil on the site is the most fundamental part of the process. Vetiver, therefore, could be a key to success in many of the different sustainable-agriculture systems under development for Third World conditions.

This and other vegetative systems of erosion control should provide long-term stability and, if combined with good crop-rotation practices, such as the use of green manures and organic mulches, could lead to stable sustainable farming that might even render slash-and-burn cultivation obsolete in many places.

Trials are urgently needed throughout the tropics.

Economic Development

Despite general opinion, vetiver hedges can provide a number of products that are especially useful to farmers in the tropics. These products can be harvested without sacrificing erosion control. They will provide farmers with extra income, and this should enhance everyone's interest in establishing vetiver hedges on their lands.

These products include:

- Forage;
- Mulch;
- Thatch;
- Mattress stuffing;
- Animal bedding; and
- Mats, baskets, and screens.

Setting up markets for local vetiver products may be one of the best incentives for inducing the mass planting of vetiver hedges throughout any neighborhood.

BASIC RESEARCH

Although vetiver has been grown in scores of countries for decades or even centuries, not a lot is known about the plant itself. Studies should be undertaken of topics such as plant morphology, physiology, ecology, and cold tolerance.

Morphology

Despite the general impression that vetiver is a single clone, it contains much variation. In one project in India, for example, six different botanical collections grown side by side were strikingly different in color, rigidity, flowering, and other features. Each type could easily be distinguished, for example, by the length and strength of the culms and by the barbs on the leaf margins.[2]

Everyone interested in vetiver should now search through the germplasm for "super vetiver" cultivars for erosion control. Attributes to look for include the following:

- Method of forming clumps;
- Persistence of clump integrity;
- Structural strength;
- Root form; and
- Level of sterility.

Physiology

Much valuable research could also be accomplished by studying details of the plant's physiology. Researchers experienced in the study

[2] Information from A.M. Krishnappa.

of maize, sorghum, or sugarcane could provide useful insight here. More information is required on the following subjects:

- Seeding variability within and among genotypes.
- Flowering. Some cultivars have flowering heads; most have none.
- Pollination.
- Structural unity. Why don't the stems rot and fall off?
- Nitrogen fixation. Does vetiver by any chance benefit from associated nitrogen-fixing bacteria as does Bahia grass and some other tropical grasses? Most C_4 grasses have associated nitrogen fixation by *Azospirillum* that live in the rhizosphere.
- Allelopathy. Does vetiver adversely affect neighboring plants?
- Silica content of leaves. Is a high silica content what repels insects?
- Mycorrhizae. Are vetiver roots colonized by the beneficial fungi called "mycorrhizae"? Can mycorrhiza inoculations benefit the plant's growth?
- Water relations. What gives the plant its ability to withstand waterlogging and even submergence on the one hand and severe drought on the other?
- Mineral nutrition.

Ecology

To understand more completely the potential for problems as this plant goes global, we need to understand its synecology (the structure, development, and distribution of vetiver communities in relation to their environments). For example:

- What is the plant's best ecological niche?
- What other plants grow with it under normal conditions?
- Which animals, microorganisms, and insects associate with it?
- What are its soil interactions?
- What are its shade and water tolerances? Growth chamber studies could be useful here.
- Is the oil in vetiver roots fungicidal? Bactericidal?

Cold Tolerance

Vetiver's gravest limitation is that it is currently restricted to the warmer parts of the world. Wherever cold occurs, even infrequently, the plant cannot be relied on to hold back soil down the years. Breaking through this cold barrier would open the possibilities of using vetiver with confidence in the frost-prone parts of the planet. This might best be done by an organized search for winter hardiness among the various genotypes.

OPERATIONAL RESEARCH

In most cases vetiver can merely be planted and left alone. Nonetheless, a detailed knowledge of the conditions for its optimum performance would be extremely beneficial. Research on this might include subjects such as:

- The best spacing between the hedges down the slopes;
- The best spacing between the plants within the hedges;
- The nutrients that, at the time of planting, ensure optimal survival and growth; and
- Relations between pruning and regrowth.

Other features of hedge establishment and maintenance worthy of investigation include:

- Mass planting techniques. The machines used to plant tobacco, beach grasses, trees, and vegetables should be tested.
- Herbicide sensitivity tests.
- Assessment of weediness.
- Potential as host for pests and diseases.

Effects on nearby crops also deserve detailed investigation. These include checking for the following:

- Nitrogen deficiency in the soil next to the hedge;
- Competition for other nutrients;
- Root interactions with neighboring plants; and
- Moisture deficiencies in the soil beside the hedges.

The integration of vetiver into various farming systems also deserves study. For example, how might vetiver hedges be fitted into alley cropping, slash-and-burn, terrace farming, horticulture, rice paddies, large-scale mechanized agriculture, and other types of farming?

RESTORING DEGRADED SITES

Degraded lands are so prevalent and so vast in the tropics and subtropics that to restore even a fraction of them may seem impossible. In this regard, vetiver unaided can do little, but it offers a starting point, or at least a ray of hope. To finally have a way to preserve whatever meager soil remains is a first step. From this, techniques for integrating various crops, trees, or wild plants that will help to rebuild the capacity of the land can then be envisaged.

Something of this sort has been pioneered in northern New Zealand where earlier this century huge barren coastal dunes were slowly rolling over valuable farmland. Nothing would grow in the sand—nothing, that is, until marram grass, *Ammophila arenaria*, was tried. This salt-tolerant European species put down deep roots, collected wind-borne sand, and stabilized the giant sandy juggernaut in just a few years. That was a feat, but the true payoff occurred after foresters planted a shrubby legume (*Lupinus arboreus*) amongst the grass. The sand then began to build up fertility (notably nitrogen) and organic matter. Within another year or two, the site was fertile enough for pine seedlings to thrive among the shrubs. The eventual result: the giant dunes were tamed; for decades the forest there has provided much of the timber used in Auckland, the country's largest city.

Marram grass is not like vetiver, but it acted as a rugged pioneer that enabled the restoration of this highly degraded site. Vetiver, with its even broader adaptability and greater potential, seems likely to do the same for many barren lands in the tropics and subtropics. This possibility should be evaluated.

SPECULATIONS

An idea as revolutionary as this one opens several possibilities for innovations, a few of which are discussed below.

Strip Farming

These days everybody thinks of farms in terms of hectares or acres. But John Greenfield suggests that in extremely arid or thin-soiled areas "a one-dimensional farm" might be run along the back of a vetiver hedge. It would be only one row wide, but might run for many kilometers. The crops would grow in the narrow band of soil trapped behind the hedge and would also be watered by the concentrated runoff. This is an unconventional notion, of course, but it just might be a way to get productive farming into areas so arid that people now consider the production of any food or tree crops impossible.

Warding Off Weeds

As mentioned earlier, vetiver is reputed to be a barrier to grassy stoloniferous weeds. This should be investigated. If it proves true, vetiver hedges would likely be planted in many places not to stop erosion but to stop the invasions of Bahia, St. Augustine, couch, kikuyu, or other pestilential grasses that constantly creep into farmers' crops on silent stolons.

Blocking Crabs and Rats

Creeping grasses are not the only pests that might be held at bay. Vetiver's dense network of oil-filled roots may block burrowing creatures of various kinds. Examples of such subterranean pests are the crabs that attack paddy bunds; and moles, mole rats, and other burrowing animals.

This, too, is an unorthodox idea, but not an idiotic one. Trials at Kericho in Kenya have already shown that mole rats cannot abide a vetiver root in their burrows. They cast it out or tunnel around it and block it off, apparently to keep the smell from their living quarters.[3]

Repelling Insects

As earlier noted, insects avoid vetiver oil and vetiver roots. This might have some modern utility. Would solid bands of vetiver block termites, fire ants, or other insidious underground insects? Might the oil or the smoke from burning roots be used against household pests and vectors of disease?

Foiling Fires

Vetiver burns in a lively fashion when the tops of the plant are dry. But in many places its roots, tapping into reserves of deep moisture, keep the plants lush and green long into the dry season. Bands of these succulent plants across hill slopes are said to make good firebreaks. This may not be a wholly reliable method of fire prevention, but many fires in rural Third World areas creep along the ground through the dry grasses. It seems possible, therefore, that stout, unbroken lines of vetiver might be a godsend to foresters and farmers who, only too often, see years of effort and millions of investment dollars go up in smoke in a single afternoon.

Eliminating Striga

One small plant is a big constraint to cereal production in Africa and India. Usually called striga or witchweed, it is a parasite that spends its first few weeks of life living off the juices of other species. Its roots bore into neighboring roots and suck out the fluids. The vegetative victims are left dried out and drained of life.

Unfortunately, striga thrives in maize, sorghum, millet, cowpeas, and other valuable crops. Millions of hectares of farmland are constantly

[3] Information from C.O. Othieno.

threatened; hundreds of thousands are infested each year. And today, nothing can be done. When striga breaks out, farmers abandon their land. Some of the most productive sites now lie idle—victims of this abominable sapsucker.

Vetiver is a member of the same subtribe of grasses as sorghum and maize, and it may prove to be a sacrificial barrier to the spread of striga. Alternately, the oils in its roots may suppress this powerful parasite that does its dirty work underground. Sucking in a dose of vetiver oil may be enough to do it in.

Catching Birds

In Zimbabwe, wildlife researchers have found that blocks of napier grass, strategically placed, can attract nightly flocks of weaverbirds. These grain-devouring pests (usually called quelea) like to roost together in the tall grass after a day in the fields. The simple concept of providing a man-made haven for the night offers a means for capturing them in quantity. On moonless nights they can be approached and either caught for food or otherwise destroyed.

Despite the promise inherent in this approach to one of the world's worst pest problems, there are at present several operational difficulties. One may well be overcome using vetiver, which would likely be an ideal grass for creating the trap roosts. Compared with napier grass, vetiver would be permanent, nonspreading, and safe from wandering wildlife or loose livestock. Given further innovation, perhaps blocks of vetiver will eventually be used as "lenses" to focus flocks of pestiferous small birds wherever they are a farm problem.[1]

OTHER SPECIES

Vetiver represents a whole new approach to erosion control. The hedge concept is a point of departure for future elaborations. We have included a section on the idea of searching for more species to use where (or if) vetiver fails. For details see Appendix B.

---

[1] More information on the pioneering work in Zimbabwe can be found in the companion report on Africa's promising native cereals, *Lost Crops of Africa Volume 1: Grains*. For information on BOSTID publications, see page 162. Napier grass is mentioned in Appendix B.

Appendix A

Great Challenges, Great Opportunities

This appendix highlights some environmental horrors and points to the role vetiver might play in alleviating them.[1] The examples are presented to stimulate exploratory trials in those parts of the world where such devastating problems exist. Whether vetiver will work in sites like these is far from certain. In some, the plant will meet its greatest difficulties in just surviving. In a few, it may survive but fail to be effective. Nonetheless, it is a testimony to the potential inherent in vetiver that a single species could even be considered for such an array of challenging situations. Also, it is a testimony to the potential inherent in their green lines of grass that solutions to so many seemingly intractable problems can even be envisioned. Where vetiver fails, alternative species that make dense hedges should be sought.

[1] It has been completed by the study's staff director and does not necessarily reflect the views of others.

In popular perception, the ultimate erosion site is one in which all the soil is stripped away and only bedrock is left. In reality, however, some form of "subsoil" normally remains (as in this case in Indonesia). The site may have lost all of what scientists call "weathered soil"—the upper part that had been exposed to air, rain, heat, cold, plant life, microbes, and the other natural forces that create a friable foundation for vegetation to grow in. But in most cases, something approximating soil remains.

This so-called subsoil is uninfluenced by biological activity, and few plants can grow in it. Vetiver is one of the few that can, at least in some cases. In fact, the grass may be the long-dreamed-of tool for getting degraded, tropical, lateritic soils (oxisols) into productive use. Even a tenuous hold on just some of these subsoils could give important benefits. Vetiver hedges backed up with other plants, especially legumes, might become a soil-recovery tool of exceptional value in these days when people fear that the world is running out of farmland. (Photo: Hugh Popenoe)

Opposite: The deep soil here could be producing a lot of food, forage, or perhaps forests, but instead is being inexorably lost. Ideally, runoff pouring from the land behind should be deflected and dispersed far above this site. (That might well be done using vetiver hedges.) However, it is likely that the landowner will opt instead to merely try to stop this gully from worsening. (Indeed, it is a measure of the expense or ineffectiveness of the better known techniques that erosion has been allowed to progress so far on this valuable property.)

For such a task, vetiver seems very suitable. Starting at the head of the gully, lines of the grass can be planted from left to right. The lines should be spaced several meters apart and each should begin at a level several meters above the bottom on one side, run across the gully, and up to the same level on the other side. Once established, these grassy barriers will check the force of the rushing runoff and stop the silt from leaving the site. Through the years, a series of stable terraces will form, eventually resulting in an almost rounded depression that can be grassed and then grazed or safely planted with trees.

This particular location in rural Australia is not in vetiver's prime geographic zone; however, countless similar scenes can be found throughout the hot regions where vetiver thrives. (Photo: Australian Academy of Science)

Overleaf: When watersheds are degraded to the point where they no longer absorb torrential rains, property can be destroyed and lives lost even at a great distance. With deforestation laying bare more and more of the world's sloping lands, flooding is an increasing hazard to everyone. Rushing runoff can pour into the lowlands with a power beyond stopping even in city streets.

Protecting vast watersheds with thousands of vetiver rows might well seem nigh on impossible, both logistically and financially. However, compared to conventional berms, bunds, and terracing, such grassy dams may be practical on a wide scale. Massive efforts would certainly be required, but this could well pay off many times over if it prevents a single incident like the one shown here in Hong Kong. Cities (or maybe insurance companies) might well find it in their interest to stimulate the necessary vetiver plantings on the faraway hills. (Photo: Hong Kong Government Information Services)

Above: In most parts of the tropics, deforested lands end up covered in weedy grasses such as the *Imperata cylindrica* (known variously as imperata, lalang, alangalang, cogon, and other names) shown here on the lower slopes. Recovering such wastelands involves two challenges: fighting the vigorous weed and ensuring that erosion will not take over once the vegetation cover is disturbed.

For decades, people have searched for ways to recoup lands like these. Vetiver could be the key. It both competes with aggressive grasses and safeguards against erosion. In the case shown, rows of vetiver along the contours could be a first step toward reclaiming the unused lower slopes. Such hedges could then make it worth a farmer's time and effort to grow either crops or trees between the rows.

Should this speculation prove out in practice, the benefits would be immense. Restoring such slopes would give millions of the poor and destitute both land and a chance for a favorable future. It might even slow down (or perhaps stop) their destruction of the forests on the higher hills.

This picture was taken in Papua New Guinea, but grassy unused lands like this can be seen on all sides throughout most of the tropics. In Asia alone, imperata is keeping more than 50 million hectares out of useful production.[2] (Photo: Noel Vietmeyer)

---

[2] A leguminous tree, *Acacia mangium*, is also showing promise for combatting imperata. Together with vetiver, it could make a two-pronged attack of great value. This tree is described in a companion report, *Mangium and Other Fast-Growing Acacias for the Humid Tropics*.

Although the idea that trees stop erosion is almost universally accepted, it is more accurate to say that forests are what really do the job. As can be seen here (in Haiti), trees by themselves can do little unless the soil around them is somehow protected. A carpet of ground-level vegetation or the litter of fallen leaves is what normally saves the site from losing its surface. Lacking such support, the trees here are obviously very vulnerable. The collapsing bank behind and probable washouts in front have doomed even these resilient leucaenas.

For much of tropical forestry, vetiver should be an ideal adjunct. Here, for instance, the young trees could be easily supported by just a few short vetiver hedges behind and perhaps one in front. (Ideally, the hedges should have been put in before the trees were planted.) Not only would the soil then be stabilized, the trees would almost certainly grow faster and better because of the moisture, leaf-litter, and silt collected behind the hedgerows. Vetiver, moreover, will protect the site whenever the soil might be re-exposed—for example, if the trees are ever defoliated by hurricanes, clear-cutting, pests, or people desperate for forage or firewood. (CARE photo by Dan Stephens)

People normally think of erosion in terms of the loss of land, but the soil is not actually lost—the wind or water just moves it to a different location. Often this process destroys as much in the new location as in the old. Throughout the world, facilities worth billions of dollars are annually being rendered useless by a deadly burden of unwanted dirt. Silt is inexorably ruining rivers, reservoirs, estuaries, ports, coral reefs, wetlands, waterways, and irrigation schemes in virtually every country that has them. This useless, soil-filled reservoir in Haiti is just one example.

Vetiver plantings in the watersheds, which may be hundreds of kilometers away, offers a chance to protect such sites. It is cheap, easy, effective, almost maintenance free, and compatible with most farming and forestry activities. With vetiver, remote watersheds no longer need be left to spew out silt.

Governments, utility companies, multinational lending organizations, and others whose interest lies in saving public works from systematic ruin by siltation should test out vetiver rapidly. The economic value inherent in extending the lifetimes of water supplies, hydroelectric generators, drainage facilities, flood-control operations, storm-water overflows, low-lying highways, and other such fundamentals of national infrastructure could be enormous. (Photo: W. Hugh Bollinger)

All over the world, millions of trees are threatened not by people but by erosion. Globally speaking, soil loss probably poses a bigger threat than loggers. In an era when trees are increasingly important to the health of the earth, tree protection should be as valued as tree planting. Where erosion is the enemy, vetiver could be vital.

The grass could save this particular tree in Yatenga Province, Burkina Faso. It would likely rejuvenate the whole site. Long hedges planted roughly across the prevailing gullies would slow the water flow and halt the loss of soil. Within one or two rainy seasons the gullies would be gone. Moreover, the dense hedges would likely benefit the tree by increasing soil moisture. Indeed, local people might voluntarily plant more trees or perhaps crops on this now all-but-unused site because the heartbreak of washouts and undermined plants would have been removed. (Photo: ©Mark Edwards/Still Pictures)

Roads commonly spawn severe erosion. The smooth, flat surface collects water and cascades it over the banks, washing out the slopes below. Slopes above the road are also erosion prone, especially as they are often extremely steep. Embankments freshly cut out of hillsides or built up from fill are particularly vulnerable. The damage can be immense—to the roads, to the neighbors, and to the environment in general.

In a few tropical countries, engineers and maintenance crews already employ vetiver to stabilize roadsides. Many more could take up the practice. Normally, the benefits are realized only over the long term, but in Malaysia nursery-grown vetivers have been planted side-by-side to give "instant" protection.

And roads are not the only man-made structures that induce costly soil losses. Parking lots, airfields, driveways, irrigation pipes, bridge abutments, posts, pylons, walls, as well as construction sites are all prime erosion incubators. For these and many others, vetiver could be a cheap and easy answer. It can survive in many disturbed subsoils. It can take the hazards of humans preoccupied with work. Even an occasional errant vehicle does it no lasting harm.

The picture here shows a logging road in the Cameroon. Remote dirt roads like this seem especially suitable for vetiver; conventional techniques are just too expensive to contemplate. (Photo: WWF/Claude Martin)

The effort required to move dirt for erosion control is enormous, and the results are often less than satisfactory. On steep slopes in high-rainfall areas, for instance, runoff water piles up in ditches like this one until it finds a weak point to break through. Once that happens the soil loss can be greater than if the slope had been left untouched: the cascading water creates a sort of domino effect that breaches terrace after terrace right down the slope. All in all, physical barriers are inherently unstable in tropical areas where annual rainfall is often measured in thousands of millimeters.

Moreover, even gentle rains wash soil down into the ditches, filling them in and erasing all the hard work. (In the case shown, the back wall is steeper than the soil's natural angle of repose and is thus highly susceptible to collapse.)

It seems probable that in thousands of sites like this one in Kenya vetiver hedges can be installed for a mere fraction of the normal expense and effort. Planting hedges across the slopes requires no movement of soil. At the very least, vetiver could be employed to strengthen this ditch. The front face could be reinforced and protected and the back wall might be planted with vetiver rows to slow down the dirt that will inevitably return. (Photo: FAO)

Although enormous effort is required to build terraces, it is all wasted unless the site is continuously maintained. In the tropics, thousands of terraces like these in Ethiopia have failed, degraded, or even washed away when they were neglected. Left to themselves, terraces can actually promote erosion, and maintaining them is becoming harder and harder. Not only it is ever more expensive, but finding people to do the tedious and backbreaking work is increasingly difficult (especially with the children off at school). Under these conditions the terraces decline, the erosion accelerates, and the farm yields decrease.

Vetiver has excellent promise for protecting the eroding faces of terraces throughout the warmer parts of the world. Moreover, it can add new flexibility. With the tough grass protecting the lip, terraces can be "forward sloping"—a style that is less susceptible to washouts and takes up less land than the current (backward-sloping) kind. The grass-covered face guards against gullying whenever the terraces are overtopped. (Photo: A. Rapp)

Watersides are major sources of both soil erosion and water pollution. Hundreds of thousands of rivers, streams, lakes, wetlands, canals, conduits, culverts, ditches, fish ponds, farm ponds, effluent ponds, "tanks," drains, swales, swamps, settling basins, reservoirs, and other kinds of waterways are affected every day. Water biting into the bottom of banks like these along the Amazon inevitably tumbles soil in. Yet waterside erosion is difficult to overcome—especially in rural Third World sites, where concrete and other engineered structures are impractical.

Vetiver, however, is at home on the interface between land and water. It is one of the few terrestrial plants able to take wet conditions, even total immersion. For this reason alone, it might become outstandingly useful.

So far this possible utility has hardly ever been tested. Although certain fish ponds in Malaysia and China have been hedged in with the grass, and although riverbank rows of vetiver are being used to save a small rural airport in Nepal, the widespread potential of erosion-control hedges along the banks of waterways is not widely appreciated. In this concept, however, there could be myriad possibilities for protecting waterways of all kinds. (Photo: © Mark Edwards/Still Pictures)

Among vetiver's many strange qualities, the ability to withstand both dryness and flooding stands out as something that might be put to practical use. Could this plant be employed in helping treat wastewater in the tropics? It certainly seems possible that grass hedges could capture sediments and thereby clean wastewater before it is released into the environment.

In various parts of the world, ponds and marshes are starting to be used to purify wastes from industry, cities, and agriculture. Scientists in both Europe and North America, for instance, are finding that an artificial wetland works at least as well as a conventional treatment facility. They employ reeds, bulrushes, and other aquatic plants. In the hot parts of the world (such as this site in Mexico) vetiver might be a superb substitute or complement.

Artificial wetlands might also be used to filter more than just sewage. Other possibilities are: process water used in industry, animal-waste runoff, pesticides leaching off farmlands, and storm water. The last is of particular concern these days. Rains wash motor oil, tar, animal droppings, and battery fluids from city streets and sluice them down what are known as "storm-water drains." A few cities are experimenting with filtering such chemical cocktails using man-made marshes. New Orleans, for example, aims to cut its storm-water pollution by half this way. (Photo: ©Mark Edwards/Still Pictures)

A common practice, recommended by authorities throughout the world, is the gully check. This is typically a "dam" of masonry, rocks, sandbags, logs, or brushwood. Thousands are installed in isolated catchments in every country.

These structures can cost a lot to build, and their installation often takes a huge amount of effort. Yet they are always temporary. Gully checks inevitably silt up, breach, or get bypassed. Typically, runoff flows over or around the barrier, digging holes and ultimately collapsing the whole structure.

Vetiver seems to offer a better solution. Unlike these "dead dams," a vetiver hedge is alive and self-adjusting. It grows, it bends, it rises up, and it anchors itself deeply in the soil. Moreover, it spreads the runoff out, and any excessive flow runs harmlessly over it and down the grass-covered face of a naturally formed terrace.

This location in the Colombian highlands should be ideal vetiver country. (Photo: D. Henrichsen)

Wherever coconuts are processed, mountains of residue remain. The husk surrounding the nut is almost impervious to decay, and although the longer fibers (known as coir) may be woven into mats and other products, the shorter ones pile up in huge, unsightly heaps. Sri Lanka, shown above, has an estimated 35 million tons of this so-called "coir dust" lying unused.

Now, however, there seems to be a breakthrough. Almost a decade's research has shown that coir dust is one of the world's best mediums for cultivating plants. It absorbs and holds moisture, it retains a light and open structure, and it resists decay and fungal diseases. European plant nurseries and flower growers are starting to use it in place of peat or sphagnum moss.

Even more potential might be found in using coir dust for rejuvenating worn-out soils in the tropics. Mixed with manure, it creates an almost perfect medium for high crop yields.

Probably this simple idea has not previously been implemented because tropical rains wash coir compost away too readily. Here is where vetiver comes in. Hedges of the dense grass could be just the ticket for holding the highly movable soil amendment in place. In the uplands of Sri Lanka, this combination is already showing promise. Vetiver hedges are holding the man-made medium in place, and crops are being grown in coir dust and manure on a barren area that is almost down to bedrock. (Photo: Lee St. Lawrence)

Even in deserts, rain often falls in torrents. In fact, periodic flooding is a paradox of many arid lands. However, all too often the soils are unable to absorb the deluge; water of immense value rushes away uselessly (and sometimes destructively) down wadis. To capture the bounty of desert downpours has been the aim of what is generally known as "water harvesting." In one popular approach barriers are erected across the wadis, gulches, gorges, and other natural funnels of flash flooding.

Rock walls, like those being erected here in West Africa, are a common form of barrier. Grassy hedges might be much better. They would be easier to install, unlikely to wash out, and would grow taller as silt builds up behind—a feat that rocks haven't learned yet. Moreover, vetiver is more than just a barrier. It is also a trickle filter that keeps excessive flows spread out so they can sink evenly into the land. All in all, a series of hedges could rob gully-washers of their force and keep the water (and the silt) right where it is most needed. Even in extremely dry zones like this one near Agades in Niger, moisture deep in the wadi bottom should keep the hedges alive through even the driest seasons. (Photo: Oxfam-America)

Vetiver may be a "safety belt" that can help protect sites endangered by desertification. Although apparently native to swamps, the plant shows remarkable ability to survive in dry lands as well. A site like the one shown is obviously a challenge, but vetiver hedges might prove capable of holding back the sand, stabilizing the dune, and saving the houses from destruction. Moisture builds up in sand dunes, and this grass, with its vast and deep root system, should tap into it well.

Assuming vetiver can survive in this dune near Niamey in Niger, hedges along the back side will slow and maybe halt the sand's forward progress. Rows across the face would also help. This simple technique certainly should be within the means (both financial and technical) of these homeowners, who in most remote and impoverished locations are the only ones likely to do anything. The villagers themselves should be enthusiastic cooperators—seeing in this natural and easy-to-understand method not only a chance to save their homes and fields but also a way to get back to producing trees or crops on the sandy slopes. (Photo: ©Mark Edwards/Still Pictures)

To tame moving sands in drought-ridden regions, the authorities sometimes set up massive gridlike networks of palm fronds. Each frond is laboriously cut from a palm and dug into the dunes. If vetiver can survive here, it would probably be more effective. This deeply rooted plant is unlikely to be blown over. It can rise up to match sand accumulations and is thus unlikely to be buried. It also resists grazing, even by goats. This grass, which can be taller than these fronds being planted in Tunisia, might prove to be practical for protecting foot tracks, roads, railways, canals, villages, houses, forests, and other facilities from blowing sand.

Further, in areas such as West Africa, vetiver's strong but resilient foliage should make an excellent windbreak. Farmers might find it especially valuable because the winds rise with the onset of the rains, and fine sand blowing across the surface of the land often buries crops before their seedlings can even get a grip on life.

Of course, the climate in this area is very different from that where vetiver is best known. Despite the moisture in the dunes, the plant is quite likely to fail. However, a vetiver relative (*Vetiveria nigritana*) is native to Africa. It, as well as other tall native grasses, may prove better. (Photo: FAO)

Appendix B
Other Potential Vetivers

To rely on a single species for hedges to control erosion throughout an area as large as the tropics is so unwise that a search for alternatives should be started quickly. This appendix identifies plants that might be employed either along with vetiver or as "safety-nets" in the event that vetiver develops problems in widespread practice. It should be understood that none of these species, as far as we know, has *all* of vetiver's properties. It should be understood also that additional species (those we haven't thought of) might be as good or even better.

With 10,000 different species of grasses worldwide, it seems logical to assume that vetiver is not the only one with the exact combination of properties required for erosion-control hedgerows. In the main, the grasses mentioned below are tall, vigorous perennials, adapted to many soils and site conditions.[1]

It is important to realize that the most readily available strains of these grasses may be the least valuable in erosion control. They are likely, for example, to have been selected for features such as palatability to animals, high seed yield, or soft stems. The types best for erosion control, on the other hand, may be the rough, tough, "unproductive" and highly unpalatable types that were previously rejected at first glance.

Also, it is worth remembering that related grass species can often be hybridized. This can produce seedless hybrids that previously may have been considered worthless. But in erosion-halting hedges, such a reproductive flaw would be an asset.

---

[1] By and large, only grasses are included in this appendix. A review of shrubs and trees for hedgerows in developing countries is already available. G. Kuchelmeister. 1989. *Hedges for Resource-poor Land Users in Developing Countries.* Deutsche Gesellschaft für Technische Zusammenarbeit (GTZ), Eschborn, Germany. 256 pp. We have, however, included several shrubs that are not mentioned in that fine book that we think may have potential.

VETIVER'S CLOSE RELATIVES

Perhaps the most promising alternatives to vetiver will be found among its own wild relatives. As has been noted already, vetiver has a dozen or so close relatives (see sidebar, page 116). These species belong to the same genus, but so far none has been explored for its erosion-controlling capabilities. Nothing suggests that any of them can match (let alone surpass) vetiver itself, but they deserve research attention nonetheless.

All these vetiver relatives are wild plants. Presumably, they are all fertile and spread by seed. However, none is known to be a weed or nuisance of any moment.

These plants are widely scattered. Four are native to Queensland, Australia, and some of those can also be found in the neighboring areas of New South Wales and Northern Territory, as well as in New Guinea. One is native to India—not to the northern plains and swamps like vetiver itself, but to the southwestern region in Karnataka. One is native to the islands of Mauritius and Rodrigues in the Indian Ocean. And two are native to the African continent.

The fact that these grasses mostly occur in riverine basins and other wet areas would seem to argue against their usefulness on hill slopes. However, vetiver itself also comes from a soggy background. By and large, they are robust plants with stout root stock and erect stems. Such features might make them good for erosion control, but the roots on at least one are said to grow horizontally, which would pose problems in hedges running across farms or forests.

VETIVER'S DISTANT RELATIVES

Lemongrass

Apart from vetiver itself, lemongrass (*Cymbopogon citratus*) has probably been used in hedges against erosion more than any other grass species. It is planted on bunds for soil conservation as well as on hillsides and road cuts in Central America and elsewhere in the tropics.

It has a number of vetiver's vital features. For example, its large and strong stems can hold back soil, even when it is planted in a single line. It has wide adaptability and the capacity to survive where terrain is difficult and conditions terrible. It tillers strongly, forming large tussocks. Because it seldom flowers, it does not spread by seedlings.

All of these are features favorable for erosion-control hedges, of course. But in practice, lemongrass does not seem to work as well as vetiver. In Costa Rica, for instance, farmers who use both are slowly abandoning lemongrass. They say that compared with vetiver, it requires more labor. They have to replant their lemongrass hedges every four years or so, and separating the planting materials from the clumps is difficult. The hedges also are less dense, less resistant to stem borers, and more unruly. Whereas vetiver rows are regular, compact, and erect, lemongrass rows are irregular, patchy, and often extremely unkempt, with leaves hanging in all directions.

Like vetiver, lemongrass is widely distributed throughout the tropics.[2] Its oil is one of the most important perfumery ingredients, widely used for scenting soaps, detergents, and other consumer products. People throughout Southeast and South Asia flavor popular drinks, soups, curries, and other foods with its leaves.

It seems possible that amongst the germplasm of this well-known grass, forms will be discovered that are well suited for use as hedges against erosion.

Citronella

Citronella grass (*Cymbopogon nardus*) is another vetiver relative that has some of the vital properties for a barrier hedge. It, too, is a robust perennial. It, too, is cultivated for its essential oil[3] and has thus been widely distributed throughout the tropics.[4] And it, too, is sterile.

For all that, however, citronella grass has not been used for erosion control—as far as we know. Now it should be tried. The type that is cultivated, the so-called "Java type," may not be the best for the purpose. A "Sri Lanka type" that had been rejected because its oil is inferior is possibly more suitable. It is exceptionally robust and resilient.

Other members of the lemongrass and citronella genus might also work in erosion control. Indeed, about 60 *Cymbopogon* species are found in the tropics and subtropics of Africa and Asia. In general, they are coarse perennials, with aromatic leaves. None has been seriously

---

[2] It is grown, for instance, in the West Indies, Guatemala, Brazil, Zaire, Tanzania, Madagascar, Indochina, Java, Sri Lanka, Burma, Mauritius, and parts of South America. Madagascar and the Comoro Islands produce most of the oil. The plant is probably native to Southeast Asia or Sri Lanka but is unknown in the wild.

[3] It is used chiefly for perfuming low-priced technical preparations such as detergents, sprays, and polishes. Its use as an insect repellent has declined as more efficient synthetic substances have become available.

[4] The plant originated in Sri Lanka, but now is grown, for instance, in Taiwan, Brazil, Sri Lanka, East Africa, Zaire, Madagascar, the Seychelles, and the West Indies. Since 1930 a considerable industry has been built up in Central America, particularly in Guatemala.

Vetiver's Close Relatives

Perhaps the most promising alternatives to vetiver will be found among its own wild relatives, discussed below.

Vetiveria elongata

This Australian grass is found around (and even within) freshwater lagoons, damp depressions, and rivers in the Northern Territory and Queensland. It also occurs in New Guinea.

Vetiveria filipes

Another Australian species, this plant is found in Queensland and New South Wales. It is characteristic of the vegetation often growing on riverbanks, sandbanks, riverine plains, high ground near creeks, *Melaleuca* swamps, dry creek beds, and depressions in open forests. It is said to be eaten freely by cattle.

This species is of special botanical interest because its morphology is intermediate between vetiver and lemongrass (see page 114). One accession was found to have a chromosome number ($2n = 40$) twice that normally found in the genus.*

Vetiveria intermedia

Yet another Queensland species, this one also likes the sandy banks of channels, growing either in the open or in partial shade.

Vetiveria pauciflora

This, the fourth Australian species, is found in the Northern Territory and Queensland. It, too, occurs on sandbanks and riverbanks.

Vetiveria rigida

This Australian species has only recently been described.

Vetiveria lawsoni

Like vetiver itself, this species is from India. However, it is native to the southwestern region. It is, for example, common in

* R.P. Celarier. 1959. Cytotaxonomy of the Andropogoneae IV. Subtribe Sorgheae. *Cytologia* 24:285–303.

the district around Dharwar in Karnataka State. The root stock is stout; the stem is erect, simple, and slender; and the internodes are very long. Such features might make the plant good for erosion control, but the root is said to grow horizontally, which would compete with nearby crops or trees.

Vetiveria arguta

This little-known species apparently occurs only in Mauritius and the neighboring island of Rodrigues in the Indian Ocean.

Vetiveria fulvibarbis

A West African member of the genus, this grass is found from Senegal to Cameroon, including Ghana, Togo, Mali, and Nigeria. It occurs mainly on flood plains and is a robust perennial that grows up to 2 m tall. Almost nothing has been reported about the plant, but it can be quite common. For instance, in Ghana it can be found on the plains between Accra and the Volta River.

Its roots and stems are (apparently) scented. People in Mali use the roots to perfume drinking water and the stems to weave mat-like wall panels called "seccos." Both roots and stems can be bought in the markets of Bamako, for example.**

Vetiveria nigritana

This grass is the main African species. It is found in most sub-Saharan areas from Senegal to Mozambique.† It is a tufted perennial, as tall as 2.5 m. It flourishes on wet soils and grows in water or at least in wet, usually swampy, ground. It occurs, for example, on alluvial floodplains in Ghana, extending along seasonal streams and channels in areas of marsh grass and tree savanna. It also remains as a relict in the low, wet spots left unused in heavily farmed areas. It is said to tolerate slightly saline soils.

Although native to tropical Africa, this species apparently also occurs in scattered locations in Sri Lanka, Thailand, Malaysia, and the Philippines.

** N. Diarra. 1977. Quelques plantes vendues sur les marches de Bamako. *Journal d'Agriculture Tropical et de Botanique Appliquée: Travaux d'Ethnobotanique et d'Ethnozoologie* 24(1).

† Nations where its occurrence has been reported include Guinea, Mauritania, Gambia, Guinea Bissau, Sierra Leone, Ivory Coast, Burkina Faso, Ghana, Togo, Dahomey, Niger, Nigeria, Gabon, Benin, Cameroon, Central African Republic, Sudan, Zimbabwe, Zaire, Malawi, Tanzania, Botswana, Namibia, and Angola.

assessed for erosion-hedge purposes, but in Senegal, *C. schoenanthus* is used more widely than vetiver for erosion control.[5]

Gamba Grass

Yet a fourth distant relative of vetiver, gamba grass (*Andropogon gayanus*), is planted on bunds as an anti-erosion measure in parts of Africa. Farmers in Niger also use grassy hedges of this species for windbreaks.

An erect tufted perennial up to 2 m high, gamba grass is native to tropical Africa and has been introduced into other tropical areas, such as Brazil, India, and Queensland, Australia. It is adapted to many types of soil, can tolerate a long dry season, and has produced good results in northern regions of Nigeria and Ghana. It persists well under grazing but becomes unpalatable after it flowers.

Those are virtues for a vetiverlike hedge, but gamba grass has severe limitations. It is "clumpy," and gaps tend to form between the plants. Also, the individual clumps tend to die in the center. In addition, the plant spreads from seed and has shallow, horizontal roots.

There are about 100 other members of the genus *Andropogon*, and some of those may prove better than gamba grass as erosion-hedge plants. They tend to be big and brawny. Some are severe weeds, but sterile forms might be found or created by hybridization.

Sugarcane Relatives

In parts of India grasses are commonly used for erosion control—not in the single-line hedge like vetiver but in blocks or bands on bunds and berms. Among the various species employed is *Saccharum moonja*. This weedy sugarcane relative is extensively used along the bunds and in gullies to check wind and water erosion in northern India. It is unpalatable to animals and is very hardy. This and other *Saccharum* species might be useful as eventual back-ups for vetiver.

Maize Relatives

The 12 known *Tripsacum* species all have stout stalks that might make them useful in hedges. Two are already showing promise.

Guatemala grass (*T. laxum*) is used to form bench terraces in the Nilgiri Hills of Tamil Nadu, India. It is said to check erosion on steep slopes very effectively.

[5] So far, vetiver has been used only experimentally there. This may quickly change, however, because the farmers generally consider that *Cymbopogon schoenanthus* is weedy and difficult to manage.

In the United States, the U.S. Department of Agriculture is testing gama grass (*T. dactyloides*) as a vegetative contour strip for field agriculture. This native, cold-tolerant species is already widely planted as a forage, and many different genotypes have been collected. These are now under evaluation for use in erosion hedges. Plants with upright habit and other desirable characteristics (notably, disease resistance) are being examined. An especially interesting possibility is the use of gama grass as a perennial grain. Even the unimproved cultivars have yielded about 25 percent as much as wheat, without the annual costs of replanting or the erosion caused by tilling bare ground.

Sorghum and Its Relatives

Certain little-known types of sorghum have such strong stems that they are used as building materials and as living stakes to support climbing crops. They are so strong that West African farmers employ them to hold up even the massive weight of yam plants. Also, these sorghums are very resistant to rot—even when dead. For months after setting seed, they continue holding up heavy yam plants, despite the heat and humidity beneath the tentlike curtain of yam foliage.

The sorghums used this way today are annuals, but perennial sorghums are known, and to combine the strong stalks and perennial habit seems well within the realm of possibility.[6]

TROPICAL GRASSES UNRELATED TO VETIVER

Napier Grass

As noted in earlier chapters, skeptics commonly criticize vetiver on the grounds that it is not "useful" enough for farmers to want to plant it. Usually, these skeptics claim that farmers will only plant a species that could feed animals. Often, they propose napier grass as a better alternative.

Napier grass (*Pennisetum purpureum*) is native to tropical Africa but is now available throughout the tropics. It is a tall, clumped perennial that in some places is planted on bunds as a windbreak and to conserve the soil. Although the dry stems or canes are used for fencing and for house walls and ceilings, it is primarily grown for fodder.

Despite the fact that it is useful for erosion control, napier grass (also known as elephant grass) cannot be used in a single line like vetiver.

---

[6] For more details see the forthcoming companion report *Lost Crops of Africa: Volume 1, Grains*.

Its stalks are too weak and the gaps between them too wide for it to stem the onrush of soil and water after tropical deluges. Moreover, its shallow, spreading roots compete with any nearby crops.

It is likely, however, that certain genotypes (possibly those rejected by the forage developers because of coarseness, woodiness, and lack of palatability) might prove to be good erosion-hedge species. Also, there is a sterile dwarf form that might prove applicable.

Napier grass has been crossed with a wild relative (*P. typhoides*) to give a sterile triploid. This is said to be produce more fodder than its parents, but it might be even more useful as a hedge for erosion control.

Other Pennisetum Grasses

Even if napier grass never works out, perhaps some of the other 80 species in the genus *Pennisetum* might have the right combination of properties. These tend to be particularly well suited for the tropics and are already in use as food, fodder, and papermaking crops. A few have been used for erosion control (for example, moya grass, *P. hohenackeri*, which is used for controlling erosion in parts of India).

Rhodes Grass

Rhodes grass (*Chloris gayana*) is sometimes proposed as a vetiver substitute. This native of southern Africa is widely cultivated for fodder. It comes in many forms, some of which withstand fires well. However, it spreads by stolons and by seed and can become a weed. It may be useful as a sod-forming species to cover and "clamp down" eroding sites, but as an erosion hedge it is unlikely to match vetiver.

Tropical Panic Grasses

In Kenya, a panic grass (*Panicum* sp.) called "kisosi" is successfully used as an erosion hedge on gentle slopes. The stalks are laid out flat in shallow ditches dug on the contour lines and covered with soil. The internodes (spaced 10 cm apart along the stalks) produce roots and shoots. Farmers regularly cut back the shoots for fodder, and this stimulates tillering and produces a living hedge sufficiently strong to hold back an accumulation of soil about 25 cm high. In this sense, the plant operates like vetiver. However, it is killed by fire and can be used only where burning is never practiced.

Calamagrostis Species

Calamagrostis argentea is being compared to vetiver in field tests near Grenoble, France. A related species, *C. festuca*, was used as an erosion-control grass by the Incas.[7] They planted bands of it above their terraces, not only to protect soil but to spread out the runoff so it would trickle evenly down the often massive terrace systems.

TEMPERATE-ZONE SPECIES

Erosion is not confined to the tropics, of course. But, as we have noted earlier, vetiver is. In terms of the worldwide erosion problem, therefore, one of the greatest discoveries would be a temperate-zone counterpart to vetiver. Several possibilities are discussed below.

Switch Grass

The U.S. Department of Agriculture is paying special attention to switch grass, *Panicum virgatum*. This native, warm-season grass is already widely planted in the eastern half of the United States as a fodder and revegetation plant, and many selected genotypes have been characterized in detail. Of particular interest is a variety called "Shelter," so-called because it was selected to enhance wildlife habitat during winter. Somewhat like vetiver in form, Shelter has stiff, erect culms that remain upright under the weight of snow and ice.

Wheatgrasses

A "vetiver for the cool zones" might be found among the wheatgrasses (*Agropyron* and *Elymus* spp.). One possibility is stream-bank wheatgrass (*A. riparium*). This tall turf-type grass is relatively low-growing, but in Montana and other parts of the United States single or double rows of it are planted across fields to reduce soil erosion. These hedges capture the blowing snow and greatly increase crop yields. Farmers employ it mainly in semiarid areas where they want the winter snow to accumulate to build up enough soil moisture for the subsequent spring and summer crops. Another species used in the same way is *A. elongatum*, a clump-type grass that can grow to over a meter high. It also retains its stems throughout the cold Montana winters.

Strips of these and of some of the other 15 wheatgrasses have also been established across expanses of rangeland to conserve snowfall and soil as well as to provide cover and food for wildlife. They work well, but if not spaced properly, they may funnel the wind, worsen the erosion, and cause the crops to ripen unevenly across the fields.

[7] Information from R. Matos.

Pampas Grass

Pampas grass (*Cortaderia argentea*) is a temperate-zone species from the southern cone of South America. It looks for all the world like vetiver. It may also function in the same way. However, it is fertile, tends to spread, and has become a serious weed in some areas.

Several related species, commonly called "toe toe," are native to New Zealand.[8] Attractive ornamentals, they are often used to beautify homes both in New Zealand and in parts of the Northern Hemisphere (the San Francisco Bay area, for instance). However, these plants are fertile and seem likely to spread slowly in fields. Perhaps sterile hybrids could be found within, or created between, these various species.

Feather Grasses

The genus *Stipa* comprises 300 species, some of which might make useful hedge species. Native to either temperate or tropical areas, they tend to be long-lived perennials whose clumps can survive two decades or more. They tolerate burning and grazing and often grow in dry areas.[9] Their stems are so woody and strong that they have been used for making paper, ropes, sails, and mats. They are not considered weedy or aggressive, although they can spread by seed. Generally, they are considered to be free of pests and diseases.

Some examples follow:

- Esparto grass (*S. tenacissima*). Mediterranean. Used locally and exported to make fine paper; also used to make cordage, sails, and mats.
- Ichu grass (*S. ichu*). South America to Mexico. A good fodder grass in arid areas.
- Black oat grass (*S. avenacea*). North America.
- Needle-and-thread grass (*S. comata*). North American prairies. An important pasture species, it is also used for revegetating mine spoils in arid parts of Wyoming.
- Porcupine grass (*S. vaseyi*). Western United States and Mexico. Stems of this species are so strong they are used to make brushes.

Jiji-Sao

This grass[10] (*Achnatherum splendens*) is native to Siberia and Central Asia, notably Inner Mongolia. It is a long-lived perennial reaching 2 m

[8] There are at least four species: *Cortaderia turbaria*, *C. richardii*, *C. splendens*, and *C. fulvida*.
[9] Their leaves often roll up under dry conditions, thereby covering the stomata and green tissues and conferring exceptional drought tolerance.
[10] Some taxonomists place *Achnatherum* species in the genus *Stipa* (in this case, *S. splendens*).

in height and surviving in a clump for up to 20 years. Although its native habitat is wet saline marshlands, it thrives in well-drained upland soils. However, it always seems to be confined to neutral or alkaline (pH 7–9.5) sites.

So far, this species has been used primarily as a fodder; however, it has been used in erosion control. Its flowering stems are so stiff and strong that brooms are made from them. Like vetiver, it has a long, tough, and fibrous rooting system, more vertical than lateral. Amazingly resilient, it is resistant to fire, drought, extreme cold (for instance, below -30°C), and even livestock. In addition to all that, it does not seem to spread (except by seed in marshlands) and is not invasive.

Of all the species for creating vetiver-like hedges in the cooler parts of the world, this seems the most promising at present.

Miscanthus

Currently, *Miscanthus sinensis* is being tested by the U.S. Department of Agriculture as a temperate-zone counterpart to vetiver. This tall species, native to Japan and other parts of the Far East, is an attractive and increasingly popular ornamental. It looks much like vetiver and forms similar types of clumps, although they are neither as dense nor as strong. It appears to be sterile; however, the clumps spread slowly outwards, and this species is considered a pesky weed in parts of Japan.

Bamboos

It seems likely that among the 1,250 species of the woody grasses known as bamboos, good hedge species can be found or bred for both temperate and tropical regions. One species (*Bambusa oldhamii*) is already used in New Zealand. A local nurseryman, Dick Endt, points out that it performs impressively both in wind protection and erosion control. He writes: "One year in Kerikeri, I saw river-flooded land covered in silt and uprooted trees, yet the bamboo held. Not only that, but silt built up behind it, and did not kill it."

Giant Reed (Spanish Cane)

The "reed" of the Bible, *Arundo donax*, has been used for 5,000 years to make the vibrating tongues that give clarinets, organs, and other pipe instruments their voice. Its stems are so strong and fibrous that they are also used for walking sticks and fishing rods as well as for making rayon and paper. The plant has been used in erosion control in the United States and Guatemala.[11] It would appear to have wider applicability as well. It is a perennial with stout stems mostly 2–6 m tall, growing from thick knotty rhizomes. It often occurs in dense colonies but makes poor forage and is only sparingly grazed or browsed, especially when anything else is available. Apparently, it produces some fertile seed.

Throughout much of Texas, this species is used in plantings along highways, culverts, stream banks, and ditches. It is considered extremely valuable for retarding erosion, and it also provides cover for wild birds and small animals.

Despite its qualities, great care must be taken when testing this species on new sites. Recently in California, it has become a severe weed that is almost impossible to control.

Ribbon Grass

Phalaris arundinacea is used to shelter sheep in Australia. It is a clump grass that is reported to form hedges. A native of temperate Europe and Asia, it is unpalatable to livestock. Sheep, for example, will not graze it out.

SHRUBS AND TREES

As noted at the beginning of this chapter, the concept of using trees as hedges against erosion has been treated in an excellent book. However, here we present a few little-known favorites of our own.

Sea Buckthorn

Sea Buckthorn (*Hippophae rhamnoides*) is a small tree or shrub of sandy coasts and shingle banks in mountains from Europe to northern China. It is very adaptable, fast growing, and can take abuse. It is readily propagated both vegetatively (from cuttings) and from seed. Although not a legume, it is a nitrogen-fixer, and there are several reports that it is good for "restoring fertility" to degraded lands.[12]

In southeastern Russia it has successfully stabilized gullies and ravines, for example. In Inner Mongolia it is used on soils containing carbonates, in loess, and in sand. In the Danube Delta it has also been used to stabilize dunes and to reclaim open-cast mining sites where

---

[11] Information from William Luce.

[12] Like casuarinas, alders and some other nitrogen-fixing species, it uses the *Frankia* symbiosis. For more information on this interesting species, contact Lee Rongsen, ICIMOD, P.O. Box 3226, Kathmandu, Nepal.

the soils are heavy clays lacking organic matter.[13] It is also excellent as a hedge to corral or control livestock.

Russians have been studying this plant for at least 50 years and have selected and developed more that 1,000 accessions—they have even hybridized it. In large measure, this has been for uses beyond erosion control. Almost everywhere, its berries are made into jellies; in France they are stewed into a sauce for meat and fish; whereas in Central Asia they are eaten with cheese and milk. The fruits are high in carotenes and vitamins, especially vitamin C. Annual production from good cultivars can run up to about 30 tons of fruit per hectare.

The prunings from the shrubs are a much appreciated fuel. They occur in abundance and have a high calorific value.

Alders

New Zealand foresters are using European alder (*Alnus viridis*) and some South American alders (*A. jorullensis* and *A. acuminata*) as hedges against erosion. These hardy, resilient, soil-improving trees have shown good growth on some dreadful sites—down to bare subsoil in several cases and on bare rock in others.[14] This has been observed especially in the Craigieburn Range on the South Island, where the alders form healthy hedges that stabilize mountain screes (cascades of stones and rocky debris). These trees, too, fix nitrogen. Others among the 35 *Alnus* species are likely to be equally good for appropriate sites.

Leucaena

A companion volume already describes the value of this fast-growing nitrogen-fixing tree.[15] Its use in hedgerows is increasingly common in tropical areas. Indeed, many people hearing about vetiver often claim that leucaena is better.[16] However, we think that each species has its own place, and that on many sites they will perform exceptionally well

---

[13] One report stated that a 1 m x 1 m planting on a saline-eroded soil with 12–45 percent slopes colonized the entire site at the rate of 1–2 m per year from root suckers, which stabilized the area by the sixth year.

[14] In certain cases, the foresters use jackhammers to drill holes to plant the seedlings. The trees' roots as well as natural weathering slowly break down the shaly rock. Somehow the trees find enough nutrients not only to survive, but to thrive. Information from Alan Nordmeyer. Forest and Range Experiment Station, P.O. Box 31-011, Christchurch, New Zealand.

[15] For information on this fast-growing tree, see the companion report, *Leucaena: Promising Forage and Tree Crop for the Tropics* (2nd edition). National Research Council, 1984. National Academy Press, Washington, D.C.

[16] Many details on this tree and its many uses are contained in the annual report, *Leucaena Research Reports* available from The Nitrogen Fixing Tree Association (NFT), P.O. Box 680, Waimanalo, Hawaii 96795, USA.

A Promising Alternative

J.J.P. van Wyk thinks he has a plant for erosion control that is even better than vetiver. It is an African grass called weeping lovegrass. This is not a big, rough-and-tough species like vetiver. It does not form thick, dense hedges that stand like sentinels across the slopes, blocking the passage of soil. Instead, weeping lovegrass (*Eragrostis curvula*) covers slopes with a perennial carpet of vegetation that acts en masse to protect the ground from rain and wind.

Weeping lovegrass is quite well known as a forage (in the dry Southwest of the United States, for instance), but van Wyk has searched through every nook and cranny of South Africa (the plant's native habitat) seeking out special genotypes with qualities for erosion control. The types he has found are truly remarkable—as amazing in their way as vetiver. They are, he says, "plants that don't know what good topsoil is. . .in fact, they don't know the difference between topsoil and no soil!"

Van Wyk has identified about 20 types that are proving excellent for reclamation. The South African government has adopted them for protecting roads, spoil dumps, and other eroding sites. The departments of Health, Water Affairs, Minerals and Energy, and Agriculture all have initiated trials or field projects. The mining industry also got involved, and these love grasses are now stabilizing gold-mine spoil—material from a mile or more underground that is basically more sterile than the Sahara.

These are mat-forming grasses. They reduce runoff, increase infiltration, cool the land, and reduce temperature fluctuations. Some are cold tolerant and can withstand temperatures down to -10°F without getting frostbitten. Most are dwarf, erect, high-seed-yielding types. Van Wyk has found types for use in the entire range of environments from 200 mm to 2,000 mm annual rainfall, from sea level to 3,000 m elevation, and from temperate to tropical areas, including both winter and summer rainfall patterns.

To keep up with the demand, van Wyk has established seven research farms and seven seed-producing farms. Weeping lovegrass stays in place and does its job for years, he says, but it is slow to establish. Van Wyk therefore uses tef (*Eragrotis tef*), an annual that is a close relative, to quickly generate grass cover.* Tef acts as a mother crop for weeping lovegrass.

* For more information on tef, see the companion report, *Lost Crops of Africa: Volume 1, Grains*. Van Wyk's address is Research Institute for Reclamation Ecology, Potchefstroom University, Potchefstroom 2520, South Africa.

together—the vetiver in the front blocking soil loss down-slope and the leucaena behind benefiting from the accumulated moisture and soil.

In some locations leucaena seems to have matched vetiver's soil-stopping abilities, but the shrubs had to be constantly cut back and maintained in the form of a thick, dense, contiguous hedge. In one trial in India where both were left almost untouched, soil losses through thick leucaena hedges were 15–16 tons per hectare, whereas the loss through (young) vetiver hedges was only 6 tons per hectare.

Asparagus

A decade ago, Thean Soo Tee began cultivating asparagus in the Mt. Kinabalu area of Sabah in his native Malaysia. He planted this Eurasian "shrub" some 1,200 m above sea level in hedges along the contours. The asparagus grew well and quickly on this irrigated land, maturing nine months after sowing. The plant's enormous root system held back the soil and saved the sloping fields from erosion.

Tee initiated this pioneering work years before the World Bank resuscitated the then moribund vetiver idea. His concern was to protect vegetable farms from erosion. With cabbage, peas, carrots, and other crops, the earth must be turned over after each harvest, exposing it to wind and water. Sabah's vegetable areas were turning into stone-strewn wastelands.

Tee recognized that because asparagus fetches a high price in the marketplace its cultivation would boost the income of local farmers. For his originality and endeavor, he earned a Rolex Award for Enterprise in 1984.[17]

Siberian Pea Shrub (Caragana)

This rugged, dense, leguminous shrub (*Çaragana arborenscens*) is used in low shelterbelts and for erosion control in some of the coldest, driest, and most desolate areas of the American Great Plains. A nitrogen-fixing species from Siberia, it grows almost anywhere, but is best adapted to sandy soils. It can be successfully grown where annual precipitation is no more than 350 mm and where winter temperatures plunge to -40°C.

[17] Mr. Tee's current address is Department of Agriculture, Bandar Seri Begawan 2059, Brunei.

Appendix C
Selected Readings

NEWSLETTER

Since 1989, the Asia Technical Department Agriculture Division of the World Bank has published the *Vetiver Newsletter.* This newsletter provides a forum for information exchange among the members of their Vetiver Information Network. Copies are available from the Asia Technical Department, Agriculture Division, The World Bank, 1818 H Street, N.W., Washington, D.C. 20433, USA.

HANDBOOK

In 1990, the World Bank published its third edition of a handbook intended primarily for fieldworkers and farmers in developing countries. This 88-page booklet, *Vetiver Grass: The Hedge Against Erosion,* is available in English, Spanish, and Portuguese from the Asia Technical Department, Agriculture Division, The World Bank, 1818 H Street, N.W., Washington, D.C. 20433, USA.

Other organizations have translated the book into Chinese, Tagalog, Cebuano, Ilongo, Lao, Thai, Gujarati, Kannada, French, and Pidgin (Papua New Guinea).

REVIEWS

Greenfield, J.C. 1989. Novel grass provides hedge against erosion. *VITA News* July:14–15.

Greenfield, J.C. 1989. *Vetiver Grass (*Vetiveria *spp.): The Ideal Plant for Vegetative Soil and Moisture Conservation.* Asia Technical Department, Agriculture Division, The World Bank, Washington, D.C.

Grimshaw, R.B. 1989. New approaches to soil conservation. *Rainfed Agriculture in Asia and the Pacific* 1(1):67–75.

Magrath, W.B. 1990. Economic analysis of off-farm soil conservation structures. Pages in 71–96 in *Watershed Development in Asia.* Technical paper no. 27, The World Bank, Washington, D.C.

Smyle, J.W. and W.B. Magrath. 1990. Vetiver grass—a hedge against erosion. Paper presented at the annual meeting of the American Society of Agronomy, October 2, 1990, San Antonio, Texas.

TECHNICAL REFERENCES

Andersen, N.H. 1970. Biogenetic implications of the antipodal sesquiterpenes of vetiver oil. *Phytochemistry* 9:145–151.

Bhatwadekar, S.V., P.R. Pednekar, and K.K. Chakravarti. 1982. A survey of sesquiterpenoids of vetiver oil. Pages 412–426 in C.K. Atal and B.M. Kapur, eds., *Cultivation and Utilization of Aromatic Plants*. Regional Research Laboratory, Council of Scientific and Industrial Research, Jammu-Tawi, India.

Celarier, R.P. 1959. Cytotaxonomy of the Andropogoneae IV. Subtribe Sorgheae. *Cytologia* 24:285–303.

Council of Scientific and Industrial Research (CSIR). 1976. Vetiveria. Pages 451–457 in *The Wealth of India (volume 10)*. Publications & Information Directorate, CSIR, New Delhi.

Gottlieb, O.R. and A. Iachan. 1951. O vetiver do Brasil. *Anais Associação Brasileiro de Química* 10:403–415.

Kammathy, R.V. 1968. Anatomy of *Vetiveria zizanioides* (L.) Nash. *Bulletin of the Botanical Survey of India* 10(3&4):283–285.

Kirtany, J.K. and S.K. Paknikan. 1971. North Indian vetiver oils: comments on chemical composition and botanical origin. *Science and Culture* 37(August):395–396.

Marinho de Moura, R., E.M. de Oliveira Régis, and A. Marinho de Moura. 1990. Reações de dez espécies de plantas, algumas produtoras de óleos essenciais, em relação ao parasitismo de *Meloidogyne incognita* raça 1 e *M. javanica*, em população mista. *Nematologia Brasileira* 14:39–44.

Nair, E.V.G., N.P. Channamma, and R.P. Kumari. 1982. Review of the work done on vetiver (*Vetiveria zizanioides* Linn.) at the Lemongrass Research Station, Odakkali. Pages 427–430 in C.K. Atal and B.M. Kapur, eds., *Cultivation and Utilization of Aromatic Plants*. Regional Research Laboratory, Council of Scientific and Industrial Research, Jammu-Tawi, India.

Parham, J.W. 1955. *The Grasses of Fiji*. Fiji Government Press, Suva.

Ramanujam, S. and S. Kumar. 1963a. Correlation studies in two populations of *Vetiveria*. *Indian Journal of Genetics and Plant Breeding* 23(1)(March):82–89.

Ramanujam, S. and S. Kumar. 1963b. Multiple criteria selection in vetiver. *Indian Journal of Genetics and Plant Breeding* 23(July):176–184.

Ramanujam, S. and S. Kumar. 1963c. Irregular meiosis associated with pollen sterility in *Vetiveria zizanioides* (Linn.) Nash. *Cytologia* 28:242–247.

Raponda-Walker, A. and R. Sillans. 1961. *Les Plantes Utiles du Gabon*. Editions Paul Lechevalier, Paris.

Robbins, S.R.J. 1982. *Selected Markets for the Essential Oils of Patchouli and Vetiver*. G167, Tropical Products Institute, London.

Sethi, K.L. and R. Gupta. 1980. Breeding for high essential oil content in khas (*Vetiveria zizanioides*) roots. *Indian Perfumer* 24(2):72–78.

Sobti, S.N. and B.L. Rao. 1977. Cultivation and scope of improvement in vetiver. Pages 319–323 in C.K. Atal and B.M. Kapur, eds., *Cultivation and Utilization of Medicinal and Aromatic Plants*. Regional Research Laboratory, Council of Scientific & Industrial Research, Jammu-Tawi, India.

Subramanya, S. and K.N.R. Sastry. 1989. *Indigenous Knowledge about the Use of "Vetiveria zizanioides" for Conserving Soil and Moisture*. Unpublished paper, State Watershed Development (SWDC), Podium Block, Visveswaraiah Centre, Bangalore 560 001, India.

Subramanya, S. and K.N.R. Sastry. 1990. Indian peasants have long used vetiver grass. *ILEIA Newsletter* March:26.

Trochain, J. 1940. Contribution a l'étude de la vegetation du Sénégal. *Memoires de l'Institut Français d'Afrique Noir.*

Yoon, P.K. 1991. "A Look-See at Vetiver Grass in Malaysia: First Progress Report." Unpublished report available from author, see Research Contacts.

Appendix D
Research Contacts

In this appendix we list people who are familiar with vetiver and, in some cases, have planting materials available. We hope this will help readers to obtain specific advice and to locate the plant within their own country. It seems likely that nonseedy vetiver can be found in practically every tropical nation, although some inquiring and exploring may be necessary to find it.[1]

It is important that people wishing to try vetiver locate planting materials in their own regions (rather than importing seed). As has been said earlier, seed will likely lead to seedy plants, and that in turn could produce badly behaved vetivers that do not stay where they are put. One of vetiver's outstanding qualities would therefore be subverted.

Further, importing living vetiver plants is a very real threat to sugarcane, maize, sorghum, and other crops. Vegetative grass material can carry any of at least 100 pests and plagues. To avoid introducing new pathogens to agricultural crops, all vetiver cuttings or rhizomes **must** bc brought in through your national plant quarantine system.

The World Bank publishes a list of people who receive the *Vetiver Newsletter*. This is a good source of up-to-date addresses of vetiver contacts. In addition, individuals have organized local "vetiver networks" in China at the Mountain Erosion Division, Institute of Mountain Disaster and Environment, Chengdu, Sichuan (Zhang Xinbao, Vetiver Network Coordinator) as well as in Thailand at the Forest Herbarium, Royal Forest Department, Phaholyothin Road, Bangkhen, Bangkok 10900 (Weerachai Nanakorn, Vetiver Network Coordinator). The members of these networks share ideas, experiences, planting

[1] In 1992 U.S. researchers have begun testing vetiver to see if DNA probes can distinguish between seedy and nonseedy types. If successful, such a "fingerprint" could be used to characterize "local" vetivers around the world both quickly and cheaply. For further information, contact the USDA Plant Genetic Resource Unit in Geneva, New York (S. Kresovich, listed below).

materials, and enthusiasm on a continuing basis. It is an approach that lends itself ideally to this grass-roots form of erosion control and could very well benefit many more nations.

The following contacts are taken largely from the *Vetiver Newletter* recipient list of mid-1992. Most of the people listed here have agreed to help readers of this report who contact them.

Argentina

Enrique L. Marmillon, Estancias del Conlara S.A., Casilla de Correo 451, 5800 Rio Cuarto (germplasm; tissue culture)

Jorge Carlos Permicone, Benito Permicone S.A., Avenida Roque S. Pena 15074, 2740 Arrecifes, Buenos Aires

Felipe R. Rivelli, Programa de Geología Ambiental, Escuela de Geología (Universidad), Las Tipas 697, 4400 Salta, Salta

Daniel A. Torregiani, Calle 28, No. 579, Department 30F, Mercedes 6600, Buenos Aires

Australia

H. Trevor Clifford, Department of Botany, University of Queensland, St. Lucia, Queensland 4067

Deryk G. Cooksley, Queensland Department of Primary Industries, PO Box 20, South Johnstone, Queensland (germplasm)

John William Copland, Australian Council on Industrial and Agricultural Research (ACIAR)/AIDAB, Locked Bag No. 40, Jakarta, Queen Victoria Terrace, Canberra, A.C.T. (trials)

Ian Garrard, Peter Houghton, and R.S. Junor, New South Wales Soil Conservation Service, PO Box 198, Chatswood, New South Wales 2067

Peter George Harrison, Berrimah Agricultural Research Centre, Department of Primary Industry and Fisheries, PO Box 79, Berrimah 0828, Darwin, Northern Territory (germplasm and trials)

David Mather, Division of Extension and Regional Operations, Department of Agriculture, PO Box 19, Kununurra, Western Australia 6743 (germplasm)

Tony O'Brien, Sharnae Research Center, PO Box 171, Kempsey, New South Wales 2440 (germplasm)

Kenneth C. Reynolds, Soil Conservation Service, Department of Conservation and Land Management, PO Box 283, Scone, New South Wales 2337 (germplasm and trials)

Brian Roberts, School of Applied Science, Darling Downs Institute of Advanced Education, PO Darling Heights, Toowoomba, Queensland 4350

B.K. Simon, Queensland Herbarium, Meiers Road, Indooroopilly, Queensland 4068 (other vetivers)

Pat Thurbon, Dairy Husbandry and Animal Breeding Branch, Department of Primary Industries, 80 Ann Street, PO Box 46, Brisbane, Queensland 4001

Paul Truong, Soil Conservation Research Branch, Department of Primary Industries, Meiers Road, Indooroopilly, Brisbane, Queensland 4068 (physiology; salt and mineral tolerances; germplasm and trials)

Leslie Watson, Taxonomy Laboratory, Research School of Biological Sciences, Australian National University, PO Box 475, Canberra, A.C.T. 2601 (grass database)

Donald Yule, Queensland Department of Primary Industries, PO Box 6014, Brisbane, Queensland 4702 (germplasm)

Bangladesh

Nazmul Alam, Bangladesh Agricultural Research Council (BARC), Farm Gate, Dhaka 1215

Edward Brand, CARE-Bangladesh, House No. 63, Road 7/A, Dhanmondi R/A, Dhaka
Zafrullah Chowdhury, Gono Shasthaya Kendra, PO Nayarhat, via Dhamrai, Dhaka 1350
Bruce Curry, Winrock International, PO Box 6083, Gulshan, Dhaka
Qazi Faruque Ahmad, Proshika-Manabik Unnayan Kendra, 5/2, Iqbal Road, Mohammadpur, Block A, Dhaka 1207
Jeffrey Pareria, Caritas Bangladesh, 2 Outer Circular Road, Dhaka 1217

Barbados

Calvin Howell, Caribbean Conservation Association, Savannah Lodge, The Garrison, St. Michael

Belgium

Marie-Anne Van Der Biest, Flemisch Aid and Development Organization (FADO), Gebr. Van Eyckstraat 27, 9000 Ghent
Mano Demeure, Socfinco, Place du Champ De Mars 2, 1050 Brussels
Pierre Galland, OXFAM-Belgique, Rue de Conseil 39, 1050 Brussels

Belize

Thomas Post, Christian Reformed World Relief Committee, PO Box 109, Corozal Town (germplasm)

Benin

Alabi Isaac Adje, Department of Agronomy, Oil Palm Research Station, PO Box 01, Pobe, Oueme
Benin Research Station, International Institute of Tropical Agriculture (IITA), BP 06-2523, Cotonou
Ramanahadray Fils, Amenagement Bassins Versants Lutte Contre Les Feux Brousse, BP 74, Natitingou (germplasm)
Deng Shangshi, Department of Rural Water Conservation and Soil Conservation, Cotonou

Bhutan

Dhanapati Dhungyel, Essential Oils Program, Department of Agriculture, Mongar, Eastern Bhutan (Indian germplasm and trials)
K. Upadhyay, Forest Management and Conservation Project, c/o FAO/UNDP, PO Box 162, Thimpu

Bolivia

Alfredo Ballerstaedt G., Subproyecto Chimore-Yapacani, Avenida Camacho, La Paz
Jorge Luis S. Ferrufino, Programa Agroquimico, Corporacion de Desarrollo de Cochabamba (CORDECO), Universidad Mayor de San Simon, Casilla 992, Cochabamba
Paulino Huanapaco Cahuaya, Centro de Capacitacion Campesina, Avenida Busch 1590, La Paz
Hernan Z. Hurtado, Ministerio de Asuntos Campesinos y Agropecuarios (MACA), La Paz
Mathieu A.J. Kuipers, S.E.A., Casilla 7265, La Paz (germplasm and trials)

Botswana

Bruce J. Hargreaves, National Herbarium, National Museum, PB 00114, Gaborone

Brazil

Elvira Maria de Oliveria Regis, Department of Agronomy, Universidade Federal Rural de Pernambuco, Ruia Dom Manoel de Medeiros s/p, Recife 52-079
J.M. Spain, Centro Internacional de Agricultura Tropical (CIAT), CPAC Planaltina, Brazilia
Michael Thung, Centro Internacional de Agricultura Tropical (CIAT), c/o CNPAF/ EMBRAPA, CP 179, Goiânia, Goiâs 74000 (germplasm and trials)
Brother Urbano, PATAC, CP 282, Campina Grande, Paraiba (germplasm and trials)

Burkina Faso

Mahamadou Kone, Oxfam, BP 489, Ouagadougou (germplasm)
Matthieu Oeudraogo, Oxfam, BP 489, Ouagadougou
Issoufou Oubda, Project Avv-Fara Poura, 01 BP 524, Ouagadougou 01 (germplasm and trials)

Burma (Myanmar)

U Tin Hla, Forest Department, Hsaywa Setyon Street, West Gyogone, Insein, Yangon
Myo Kywe, Department of Agronomy, Institute of Agriculture, Yezin University, Pyinmana

Burundi

Bernard L. Deline, Institut des Sciences Agronomiques du Burundi (ISABU), BP 1540, Bujumbura, (germplasm and trials)
Nkurunziza Francois, Institut des Sciences Agronomiques du Burundi (ISABU), BP 795, Bujumbura

Cameroon

Enoch N. Tanyi, IRA/ICRAF Agroforestry Project, BP 4230, Yaounde (germplasm)

Canada

Malcolm Black, Conservation Service-Canada, Motherwell Building, 19001 Victoria Avenue, Regina, Saskatchewan S4P OR5
John Caraberis, PO Box 370, Pugwash, Nova Scotia, BOK 1LO
Timothy J. Johnson, Institute for Study and Application of Integrated Development (ISAID), 132 Allan Street, Suite 5, Oakville, Ontario L6J 3N5
Colin and Susan McLoughlin, Aquilex Development Inc., 1135 Connaught Drive, Vancouver, British Colombia V6H 2G9 (germplasm and trials; cold tolerance)

Cape Verde

Frances Harris, HDRA/INIA Project, Instituto Nacional De Investigao Agraria (INIA), CP 84, Praia (germplasm and trials)

Chile

D. Gabriel Banfi D., Fundo El Gardal, Puchuncaví, Casilla 8130-2, Viña del Mar, Chile (germplasm and trials)
Carlos A. Irarrazabal, Don Carlos 3171-B, Los Condes, Santiago (germplasm)
Guillermo A. N. Rueda, Programa Nacional Manejo De Cuencas, Avenida Bulnes 259, Departamento 506, Santiago (germplasm)
Alejandro M. Yutronic, Empresa Construción. Fernando Mimica Sambuk, La Lengua 01440, Punta Arenas

China

Ding Guanming, 4 Qiaoting Shangjiaojing, Fuzhou, Fujian (germplasm and trials)
Ding Zemin, Department of Rural Water Conservation and Soil Conservation, Ministry of Water Resources, PO Box 2905, Beijing
Jiang Ping, General Forestry Station, Guizhou Forestry Department, No. 48 Yanan Zhou Lu, Guiyang 550001, Guizhou (germplasm and trials)
Trevor King, Australian Council on Industrial and Agricultural Research (ACIAR), c/o Australian Embassy, Fourth Floor, Office Building, East Lakes Complex, 35 Dongzhimenwai Dajie, Beijing
Chun-Yen Kuy, Department of Plant Physiology, Testing Station, Wu Hwa County, South China Institute of Botany, Academia Sinica, Guangdong Province (germplasm and trials)
Liao Baowen, Research Institute of Tropical Forestry, Chinese Academy of Forestry, Longdong, Guangzhou, Guangdong (germplasm and trials)
Qin Fengzhu, Ministry of Forestry, Hepingli, Beijing
Qiu Jiye, Agriculture Bureau of Zhejiang Province, 63 Hua Jia Chi, Hangzhou, Zhejiang (germplasm)
Wang Xiuhao, Jiangxi Agricultural Development Corporation, No. 75, Hongdu Street, Nanchang, Jiangxi (germplasm and trials)
Xue-ming Wang, Jiangxi Province Bureau of Agriculture, Animal Husbandry, and Fisheries, Beijing Road, Nanchang, Jiangxi (germplasm and trials)
Wang Yaozhong, Water and Soil Conservation, Sichuan Provincial Department of Water Conservation, 7 Bingshang Street, Chengdu, Sichuan
Wang Zi Song, Agricultural Foreign Capital Office, Huang Hua Shan, Tanchen, Jianyang, Fujian (germplasm and trials)
Xiang Yuzhang, Division of Soil and Water Conservation, Ministry of Water Resources, PO Box 2905, Beijing (germplasm and trials)
Xu Daiping, Research Institute of Tropical Forestry, Chinese Academy of Forestry, Longdong, Guangzhou, Guangdong (germplasm and trials)
Feng-yu Zhang, Jiangxi Provincial People's Government, Beijing Road, Nanchang, Jiangxi (germplasm and trials)
Zhang Daquan, Bureau of Soil and Water Conservation, 151, Western Road One, Xian, Shaanxi (germplasm and trials)
Zhang Guang-ming, China Red Soils Development Project Joint Office, 7 Building Bei Li-Baijia Zhuang, Beijing (germplasm and trials)
Zhang Xinbao, Mountain Erosion Division, Institute of Mountain Disaster and Environment, Chengdu, Sichuan (China Vetiver Network Coordinator)
Zheng Dezhang, Research Institute of Tropical Forestry, Chinese Academy of Forestry, Longdong, Guangzhou, Guangdong (germplasm and trials)
Zheng Songfa, Research Institute of Tropical Forestry, Chinese Academy of Forestry, Longdong, Guangzhou, Guangdong (germplasm and trials)

Colombia

Douglas Laing, Centro Internacional de Agricultura Tropical (CIAT), AA67-13, Valle, Cali (germplasm and trials)

Werner Moosbrugger, Checua Project, German Agency for Technical Cooperation GTZ/CAR, AA 100409, Bogota 10 (germplasm and trials)
Karl Mueller-Saemann, Centro Internacional de Agricultura Tropical (CIAT), AA 6713, Cali, Valle (germplasm and trials)

Comoros Islands

Lafrechoux Didier, Training and Extension Unit, VANNA Project, CARE Anjouan, BP 303, Mutsamudu
Francesco Piccolo, Project Nord-Est Anjouan, BP 346, Mutsamudu

Costa Rica

Pedro J. Argel and L. Harlan Davis, Instituto Interamericano de Ciencias Agrícolas (IICA)/Centro Internacional de Agricultura Tropical (CIAT), Apartado 55-2200, Coronado, San José (germplasm and trials)
C. Buford Briscoe, La Suiza, 7151 Turrialba (germplasm and trials)
Gerardo Budowski, University for Peace, Apartado 199-1250, Escazu
Jorge León, Apartado 480, San Pedro, Monte de Oca, San José
Rafael A. Ocampo, Centro Agronómica Tropical de Investigación y Enseñanza (CATIE), Casilla 99, Turrialba 7170

Cote d'Ivoire

Jeannine Bugain, Comite internacionale des fesses Africaines pour le développement, 01 BP 5147, Abidjan 01
Peter Matron, West Africa Rice Development Association (WARDA), 01 BP 2551, Bouaké
Paul Perrault, Winrock International, 08 BP 1603, Abidjan 08
Eugene Terry, West Africa Rice Development Association (WARDA), 01 BP 2551, Bouaké

Cyprus

Luigi Guarino, International Plant Genetic Resources Institute (IPGRI, formerly IBPGR), c/o Agricultural Research Institute, PO Box 2016, Nicosia

Denmark

Thure Dupont, Svanemollevej 54, 2100 Copenhagen (germplasm)
Ole Hans Christian Olsen, Danagro Adviser A/S, Granskoven 8, 2600 Glostrup (germplasm and trials)

Dominica

Michael Didier, Dominica Banana Marketing Corporation, PO Box 24, Charles Avenue, Goodwill

Dominican Republic

Jorge L. Armenta-Soto, Caribbean Rice Improvement Network (CRIN), Centro Internacional de Agricultura Tropical (CIAT), c/o Instituto Interamericano de Ciencias Agrícolas (IICA) Apartado 711, Santo Domingo

Joachim Boehnert, Proyecto SEA/DED de Conservacion de Suelos Sistemas Agroforestales, Apartado 34, Mao, Valverde
Eduardo Latorre, Fundacion Dominicana de Desarrollo, Apartado 857, Santo Domingo
Lionel Robineau, ENDA-Caribe, Apartado 21000, Huacal, Santo Domingo

Ecuador

Ciro G. Cazar Noboa, Centro de Estudios y Accion Social (CEAS), Apartado 242, Riobamba, Chimborazo (germplasm)
Michael Hermann, Centro Internacional de la Papa (CIP), Moreno Bellido 127 y Amazonas, Casilla 17-16-129-CEQ, Quito
Miguel A. Jaramillo, La Pradera, KMT 32 Via A La Costa, Guayaquil
Carlos C. Nieto, Instituto Nacional de Investigaciones Agropecuarias (INIAP), Estacion Experimental Santa Catalina, CP 340, Quito (germplasm)
Osvaldo Paladines, Fundacion Para El Desarrollo Agropecuario Fundagro, Casilla 17-16-219, Quito

Egypt

Abdhul El-Aziz Saad, Department of Soil Science, Ain Shams University, Cairo

El Salvador

Cecilio Antonio Cortez, Final Avenida FCO Menedez, 54R #1-2, Ahuachapan
Rodolfo Antonio Olivares M., PRODAGRO, Calle Gabriela Mistral Pje. Colon, No. 13, San Salvador

England

Geoffrey P. Chapman, Wye College, University of London, Wye, Near Ashford, Kent TN25 5AH
Seamus Cleary, Catholic Fund for Overseas Development (CAFOD), 2 Garden Close, Stockwell Road, London SW9 9TY
P.J.C. Harris, Henry Doubleday Research Association, National Centre for Organic Gardening, Ryton-on-Dunsmore, Coventry CV8 3LG
R.M. Jarrold, MASDAR (UK) Ltd., 141 Nine Mile Ride, Finchampstead, Wokingham, Berkshire RG11 4HY (tissue culture germplasm)
R.P.C. Morgan, Silsoe College, Silsoe, Bedford MK4 5DT
Steve A. Renvoize, Royal Botanic Gardens, Kew, Richmond, Surrey TW9 3AB (taxonomy)
Gill Shepard, Social Forestry Network, Overseas Development Institute, Regent's College, Inner Circle, Regent's Park, London NW1 4NS
Brian Sims, Latin American Section, Silsoe Research Institute, Wrest Park, Silsoe, Bedford MK45 4HS

Ethiopia

Bebeka Project, Coffee Plantation Development Enterprise, PO Box 6, Gimira, Keffa (germplasm and trials)
Debela Dinka, Arsi Regional Agricultural Development Department, Ministry of Agriculture, PO Box 388, Asella, Arsi (germplasm and trials)
Graham Garrod, Shewa PADEP VI Project, PO Box 267, Addis Ababa
Goma One Coffee Farm, Limu Coffee Plantation Development Enterprise, PO Box 87, Ilubabor, Limu/Agaro (germplasm and trials)

Goma Two Coffee Farm, Limu Coffee Plantation Development Enterprise, PO Box 238, Ilubabor, Limu/Agaro (germplasm and trials)
Gumer Coffee Farm, Coffee Plantation Development Enterprise, PO Box 47, Kossa/Genet (germplasm)
Kossa Coffee Farm, Coffee Plantation Development Enterprise, Kossa/Genet (germplasm)
Suntu Coffee Farm, Coffee Plantation Development Enterprise, PO Box 18, Kossa/Genet (germplasm)
P. Ramanagowda, Institute of Agricultural Research, Sinoma Research Centre, PO Box 208, Bale, Robe
John Walsh, International Livestock Center for Africa (ILCA), PO Box 5689, Addis Ababa
Meseret Wondimu, Office of Adaptive Research, PCDPID, Ministry of Coffee and Tea Development, Addis Ababa

Fiji

Jale Baba, c/o—Fiji Pine Commission, PO Box 521, Lautoka (germplasm and trials)
Peter Drysdale, 25 Phlugerr, Lautoka (germplasm and trials)
Jai Gawander and Leon Sugrim, Fiji Sugar Corporation Limited, PO Box 3560, Lautoka (germplasm and trials)
Wieland Kunzel, Fiji-German Forestry Project, Box 14041, Suva (germplasm)
Buresova Nemani, c/o M.P.I., PO Box 358, Suva (germplasm and trials)

Finland

Seppo Hamalainen, Silvestria, Lapinlahdenkatu 14 B, 00180 Helsinki
Heikki Rissanen, FINNIDA, Mannerheimintie 15 C, 00260 Helsinki

France

Jacques Barrau, Laboratoire d'Ethnobotanique-Biogéographie, Muséum National d'Histoire Naturelle, 57, rue Cuvier, Paris 75231
Francoise Dinger, Centre national du machinisme agricole du génie rural des eaux et des forêts (CEMAGREF), Grenoble Regional Centre Dom. Universitaire, 2 rue de la Papeterie BP 76, Cedex St. Martin-d'Heres 38402 (germplasm)
Philippe Girard, Centre Technique Forestier Tropical (CTFT), 45 bis avenue de la Belle Gabrielle, 94736 Nogent sur Marne
Francois V. Rognon, Département du Centre de Coopération Internationale en Recherche Agronomique pour le Développement (CIRAD), 42-Rue Schaeffer 75116, Paris
Bertrand Schneider, The Club of Rome, 17 rue Camille Pelletan, 92290 Chatenay Malabry

Gambia

John Fye and Kabir S. Sonko, Soil and Water Conservation Unit, Department of Agricultural Services, Yundum Agricultural Station, Yundum

Germany

Wolfram Fischer, Division 404, German Agency for Technical Cooperation (GTZ), Dag-Hammarskjold-Weg 1-2, 6236 Eschborn
Bernward Geier, IFOAM General Secretariat, c/o Okozentrum Imsback, 6695 Tholey-Theley

GITEC Consult GMBH, Bongardstrasse 3, PO Box 320446, 4000 Dusseldorf 30
Ernst Klimm, Klimm & Partner, Forststrasse 7, 5020 Frechen 4
Klaus Schmitt, Society for the Promotion of Agriculture and Environmental Conservation (SPAEC), Langgasse 24/H, 6200 Wiesbaden-1

Ghana

Nana Konadu Agyeman-Rawlings, 31st December Women's Movement, PO Box 065, Osu, Accra
S. Yaovi Dotse, Department of Crop Services, Ministry of Agriculture, PO Box 14, Tamale (germplasm)
A.K. Nantwi, Oil Palm Research Institute, PO Box 74, Kusi-Kade
D.Y. Owusu, Ghana Rural Reconstruction Movement-Yensi Centre, PO Box 14, Mampong, Akwapim

Guatemala

Mark A. Wilson, Asociación Share, 5a. Avenida 8-07, Edificio Real Reforma 13B, zona 10, Guatemala City (germplasm and trials)

Haiti

Clifford Bellande, CARE-Haiti, BP 773, Port-au-Prince
IICA/CRIN, c/o Instituto Interamericano de Ciencias Agrícolas (IICA), lere Impasse Lavaud, No. 14, Port-au-Prince
Victor A. Wynne, Haiti Seed Store, Wynne Farm, BP 15146, Petionville (germplasm)

Honduras

Jaime E. Ríos, Gnomos Farm, AP 46, Sigvatepeque
Carlos Valderrama, Winrock International, AP 1764, San Pedro Sula Cortes

Hong Kong

Ronald D. Hill, Department of Geography, University of Hong Kong (germplasm and trials)

India

I.P. Abrol, Department of Soils, Agronomy, and Agroforestry, Indian Council of Agricultural Research (ICAR), Dr. Rajendra Prasad Road, Krishi Bhawan, 110001 New Delhi (germplasm and trials)
Agricultural Research Station, Banswara, Rajasthan
Agricultural Research Station, Fathepur, Rajasthan
Agricultural Research Station, Jalore, Rajasthan
Agricultural Research Station, Kota, Rajasthan
Agricultural Research Station, Navagaon, Alwar, Rajasthan
Agricultural Research Station, Punjab Agricultural University, Bhatinda, Punjab
Agricultural Research Station, Sriganganagar, Rajasthan
Jim Alexander and Oktay Yenal, World Bank Resident Mission, PO Box 416, 110011 New Delhi (germplasm and trials)
P.W. Amin, Punjabrao Krishi Vidyapeeth, Krishinagar, Akola, 444104 Maharashtra

S. Ananda, Department of Technology Transfer, Karnataka Welfare Society, PO Box 28, Chikballapu, 562101 Karnataka (germplasm and trials)
M. Aravindakshan, Kerala Agricultural University, District Trichur, Vellanikkara, 680654 Kerala
A.K. Aron, Kandi Watershed and Area Development, Government of Punjab, SCO No. 2449-50, Sector 22-C, Chandigarh, 160022 Punjab
Govind Marotrao Bharad, Department of Agronomy and Watershed Development, Punjabrao Krishi Vidyapeeth (PKV) Agricultural University, Krishinagar, Akola, 444104 Maharashtra (germplasm and trials)
D.D. Bharamagoudra, Yelavatti, Taluk Shirabatti, District Dharwad, 582117 Karnataka (germplasm and trials)
Ashok Bhatia, Gujarat State Land Development Corporation Ltd., 78, Pankaj Society, Vasana, Ahmedabad, 380007 Gujarat
Department of Botany, Banaras Hindu University, Varanasi, 221005 Uttar Pradesh (vetiver ecology)
Patrick Carey, CARE-India, B-28 Greater Kailash-I, 110048 New Delhi
S.S. Chakraborty, Ramakrishna Mission Lokasiksha Parishad, PO Narendrapur, District 240 Parganas (South), 743508 West Bengal
Robert Chambers, Administrative Staff College of India, Bella Vista, Hyderabad, 500049 Andhra Pradesh
S.S. Chitwadgi, Bharat Forestry Consultancy, 156/A Indrapuri, Bhopal, 462021, Madhya Pradesh (germplasm and trials)
P.N. Chowdary, Watershed Development Team, Department of Agriculture, 2-2-1130/19-5, Sivan Road, New Nallakunta, Hyderabad, 500044 Andhra Pradesh (germplasm and trials)
A.L. Cogle, Soil Division, International Crops Research Institute for the Semi-Arid Tropics (ICRISAT), Patancheru PO, 502324 Andhra Pradesh
Sham Pa Daitota, Sadashaya Prakashana, Panaje, 574259 Karnataka (germplasm and trials)
R.P. Dange and H.S. Lohan, Kandi Project, I.W.D.P. Hills, Office of Forestry, Department of Agriculture, S.C.O. No. 95-97 Sector 17-D, Chandigarh, Haryana (germplasm and trials)
S.N. Das, Orissa University of Agriculture and Technology, Bhubaneswar, 751003 Orissa
S.N. Desai, Mahatama Phule Agricultural University, District Ahmednaagar, 413722 Gujarat
P.A. Deshmukh, Marathwada Agricultural University, Parbhani, 431401 Maharashtra
P.C. Doloi, Assam Agricultural University, Jorhat, 785013 Assam
Johannes M.M. Engels, Regional Office for South and Southeast Asia, International Plant Genetic Resources Institute (IPGRI, formerly IBPGR), c/o Pusa Campus, 110012 New Delhi
A.L. Fonseca, Delhi Jesuit Society, St. Xavier's, 4 Raj Nivas Marg, New Delhi 110054 (germplasm and trials)
Konaje Gopalakrishna, Department of Agriculture, 1666 9th Main Road, Wal III Stage, Bangalore, 560008 Karnataka, (germplasm and trials)
Bharama, Devendra, and Dyamana Gouda, Dharitri Farm, Yelavatti, Chirahatti, Dharwad, 582117 Karnataka (germplasm and organic trials)
Rajendra Gupta, Medicinal and Aromatic Plants (M&AP), National Bureau of Plant Genetic Resources (NBGR), Indian Council of Agricultural Research (ICAR), Pusa Campus, 110012 New Delhi (germplasm and breeding)
T.K. Gupta, Bidhan Chandra Krishi Vishwa Vidyalaya, PO Box Kalyani, Mohanpur District Nadia, 741252 West Bengal
Shinde Subhash Hanumantrao, Department of Agronomy, Mahatma Phule Krishi Vidyapeeth (MPKV) University, Rahuri, 413722 Maharashtra
B.R. Hegde, College of Agriculture, University of Agricultural Sciences, Hebbal, Bangalore, 560065 Karnataka
Narayan Ganapa Hegde, Agroforestry Division, BAIF Development Research Foundation, 'Kamdhenu', Senapati Bapat Road, Pune, Maharashtra (germplasm and trials)

R.K. Hegde, University of Agricultural Sciences, Dharwad, Karnataka (germplasm and trials)
M.A. Hussain, Andhra Seeds Corporation, PO Box 135, Hyderabad, 500001 Andhra Pradesh (germplasm and trials)
Bhaskar Bandu Jadhar, Department of Agroforestry, Konkan Krishi Vidyapeeth, Dapoli, Maharashtra (germplasm and trials)
Mohammed Abdul Jaleel, Office of Soil Conservation, Department of Agriculture, Mahbubnagar, 509001 Andhra Pradesh (germplasm and trials)
P.N. Jha, Rajendra Agricultural University, Pusa—Samastipur, 848125 Bihar
V.K. Kachroo, Directorate of Agriculture, Srinagar, Kashmir
P. Kandaswamy, Water Technology Centre, Tamil Nadu Agricultural University, Lawley Road, Coimbatore, 641003 Tamil Nadu (germplasm)
S.P.S. Karwasara, Haryana Agricultural University, Hissar, 123001 Haryana
A.A. Khan, Birsa Agricultural University, Kanke, Ranchi, 834006 Bihar
M.R. Khajuria, Regional Agricultural Research Station, Department of Plant Breeding, Sher-e-Kashmir University of Agricultural Science and Technology, Rajouri, 185131 Jammu and Kashmir (germplasm and trials)
T.N. Khoshoo, Tata Energy Research Institute, 7 Jor Bagh, 110003 New Delhi
K. Krishnamurthy, University of Agricultural Sciences, BP No. 2477, Hebbal, Bangalore, 560065 Karnataka (germplasm)
Janaky Krishnan, Department of Botany, Sarah Tucker College, Palayamkottai, V.O.C. District, Tamil Nadu
A.M. Krishnappa, Operational Research Project, University of Agricultural Sciences, Hebbal, Bangalore, 560065 Karnataka, (germplasm)
Jha. Mihir Kumar, Pradip Smriti Sansthan, L-40 Road No. 20, Srikrishna Nagar, Patna, 800001 Bihar (germplasm and trials)
M.G. Lande, Watershed Development and Research, Ministry of Agriculture, Government of India, Krishi Bhawan Road No. 247-B, New Delhi (germplasm and trials)
Lemongrass Research Station, Kerala Agricultural University, Odakkali, Ernakulan District, Kerala (germplasm collection and oil research)
David L. Madden, U.S. Embassy, Chankyapuri, 110021 New Delhi
S.V. Majgaonkar, Konkan Krishi Vidyapeeth, District Ratnagiri, Dapoli, 415712 Maharashtra
M.M.A. Mashady, Office of Forestry, Masheshwaram Project, Vidyanagar, Hyderabad, Andhra Pradesh (germplasm and trials)
S.C. Modgal, Govind Ballabh Pant University of Agriculture and Technology, Pantnagar, District Nainital, 263145 Uttar Pradesh
P.G. Moghe, Punjabrao Krishi Vidyapeeth, Krishinagar, Akola, 444104 Maharashtra
David Mosse, OX/FAM-South India (Trust), 3, Chelvaroya Mudaliar Road, PB 541, Frazer Town, 560005 Bangalore
H. Nagaraj, At Post Bedasgaon, Taluk Mundagod, 581346 Karnataka (germplasm and trials)
Tara Nagesh, MAMEOS, Soumya-Soudha, Tirthahally Tq. Shimogadt, Devangi Post, 577415 Karnataka (germplasm and trials)
G.C. Naik, Water Works Road, District Puri, Orissa (germplasm and trials)
K.S. Nanpuri, Punjab Agricultural University, Ludhiana, 141004 Punjab
Paladi Laxmi Narayana, Maheshwaram Watershed Development Project, Department of Agriculture, PO Box Tukkuguda Viamankhal, Andhra Pradesh (germplasm and trials)
National Agricultural Research Programme, Punjab Agricultural University, Ludhiana, 141004 Punjab
National Botanical Research Institute (NBRI), Rana Pratap Marg, Lucknow, 226001 Uttar Pradesh
D.P. Nema, Jawaharlal Nehru Krishi Vishwa Vidyalaya, Jabalpur, 482004 Madhya Pradesh
I.C. Patel, Gujarat Agricultural University, Shahibagh, Ahmedabad, 380004 Gujarat
R.S. Patil, 'Dharitri,' Shiva, behind Prasad Lodge, Gadag, Karnataka (germplasm and trials)

V.G. Patil, Rainfed Watershed Project, Manoli Watershed, Government of Maharashtra, Opp. Bank of Baharashtra, Jattar Peth, Akola, 444005 Maharashtra (germplasm and trials)
V.H. Patil, Agriculture University Regional Agricultural Research Station, PO Karjat, Raigad, 410201 Karjat
Winin Pereira, 79 Carter Road, Bandra, Bombay, 400050 Maharashtra (germplasm and trials)
Bala Prasad, H. No. 24-84/56 Indiranagar Colony Lothmunta, Tirumalgiri-Secundarabad, 500015 Andhra Pradesh (germplasm and trials)
A.G. Raddi, Forestry Project Coordination Cell, New PMT Building, Swargate, Pune, 411042 Maharashtra
A.K. Rai, Indira Vikas Samity, Indira Path, Hindo, 834002 Bihar (germplasm and trials)
G.S. Rajendra, Rajya Raitha Sangha, Tirthahalli Jaluk Branch, Shimoga Dt., Komandur, 577422 Tamil Nadu
Rajendra Agricultural University, NE Alluvial Plain Zone Regional Research Station, Agwanpur, Bihar
Rajendra Agricultural University, NW Alluvial Plain Zone Regional Research Station, Madhopur, West Champarun, Bihar
Rajendra Agricultural University, South Bihar Alluvial Plain Zone Regional Research Station, Sabour, Bihar
A. Padma Raju, Regional Agricultural Research Station, Department of Soil Science, Andhra Pradesh Agricultural University, Karimnagar, Jagtial, 505327 Andhra Pradesh (germplasm and trials)
R.S. Rana, National Bureau of Plant Genetic Resources, Pusa Campus, Near NSC Beej Bhawan, 110012 New Delhi
Adusumilli Narayana Rao, Andhra Pradesh State Council of Science and Technology, 38 Nagarjuna Hills, Punjagutia, Hyderabad, 500482 Andhra Pradesh (germplasm and trials)
C.S. Rao and N.S. Rao, Andhra Pradesh Agricultural University, Regional Agricultural Research Station, 1-1/103 Jammigadda, Jagtial, 505327 Andhra Pradesh
M. Singa Rao, Soil Physical Conditions Improvement Project, Agricultural Research Institute, Andhra Pradesh Agricultural University, Rajendranagar, Hyderabad, 500030 Andhra Pradesh (germplasm and trials)
Amulya K.N. Reddy, Department of Chemistry, India Institute of Science, Bangalore, 560012 Karnataka
Narasa N. Reddy, Canara Bank, Agricultural Consulting Service, 51, Thaper House, 1st Cross, J.C. Road, Bangalore, 560027 Karnataka (germplasm and trials)
Regional Agricultural Research Station, Gassaigaon, Assam
Regional Agricultural Research Station, Jawaharlal Nehru Krishi Vishwa Vidyalaya, Morena, Madhya Pradesh
Regional Agricultural Research Station, Lam (Guntur), Andhra Pradesh
Regional Agricultural Research Station, Nandyal, Andhra Pradesh
Regional Agricultural Research Station, North Lakhimpur, 737001 Assam
Regional Agricultural Research Station, Palem, Andhra Pradesh
Regional Agricultural Research Station, Shillongani, PB 33, Navgaon, 782001 Assam
Regional Agricultural Research Station, Tamil Nadu Agricultural University, Vridhachalam, Tamil Nadu
Regional Agricultural Research Station, Tirupathi, Andhra Pradesh
Regional Agricultural Research Station, Upper Brahamputra Valley Zone, Titabar, Assam
Regional Research Station, Basuli (Gorakhpur), Uttar Pradesh
Regional Research Station, Bymore Plateau and Satpura Hill Zone, Iwa, Madhya Pradesh
Regional Research Station, Eastern Ghat High Land Zone, Semiliguda, Orissa
Regional Research Station, Eastern Zone, Haryana Agricultural University, Karnal, Haryana
Regional Research Station, Nadurai Road, Aruppukottai, Tamil Nadu
Regional Research Station, Hukumchand Basia's Building-Nard No. 13, Tikamgarh, Madhya Pradesh

Regional Research Station, Orissa University of Agriculture and Technology, Mahisapat, Orissa
Regional Research Station, Orissa University of Agriculture and Technology, Bhubaneswar, Orissa
Regional Research Station, Orissa University of Agriculture and Technology, Post Ranital, Orissa
Regional Research Station, Punhabrao Krishi Vidyapeeth, Punjab Agricultural University, Gurdasiur, Punjab
Regional Research Station, Punjab Agricultural University, Kandi, District Gurdaspur, Punjab
Regional Research Station, Tamil Nadu Agricultural University, Paiyur, Tamil Nadu
Anna Runeborg, Swedish International Development Authority (SIDA), c/o Swedish Embassy, Nyaya Marg, Chanakyapuri, 110021 New Delhi (germplasm and trials)
G.S. Sachdev, Ministry of Agriculture, Vindhyachal Bhavan, Bhopal, 462003 Madhya Pradesh (germplasm and trials)
J.S. Samra, Research Centre, Central Soil and Water Conseration Research and Training Institute, Block 3-A, Sector 27-A, Madhya Marg, Chandigarh, Union Territory
R.N. Saran, R.A.K. College of Agriculture, Sehore, 466001 Madhya Pradesh (germplasm and trials)
K.N. Ranganatha Sastry, State Watershed Development Commission, Podium Block, 3rd Floor, Visvesvarya Centre, Government of Karnataka, Dr. Ambedkar Road, Bangalore, 560001 Karnataka (indigenous knowledege)
S.N. Saxena, Agricultural Experiment Station Durgapura, Rajasthan Agricultural University, Jaipur, 822562 Rajasthan
David N. Sen, Laboratory of Plant Ecology, Department of Botany, University of Jodhpur, 342001 Rajasthan (saline agriculture; oils)
Sohan Lac Seth, Watershed Development Council, Ministry of Agriculture, Government of India, Rajendra Prasad Road, New Delhi
Parmesh Shah, Aga Khan Rural Support Programme, Choice Premises, Swastik Cross Road, Navrangpura, Ahmedabad, 380009 Gujarat
R.C. Sharma, Department of Agriculture, Land Development Corporation Limited, Hangallya Complex Nr. Vasna, Ahmedabad, Gujrat (germplasm and trials)
A. Shastri, Watershed Development Project, 2-3-1130/19-5 Vidyonagar, Hyderabad, Andhra Pradesh (germplasm and trials)
Gurmel Singh, Central Soil and Water Conservation, Research and Training Institute, 218 Kanlagarl Road, Dehra Dun, 248195 Uttar Pradesh
H.G. Singh, Sher-E-Kashmir University of Agricultural Science and Technology, 45-B, Gandhinagar-Jammu, Shalimar Bag, Srinagar, Jammu and Kashmir
R.K. Singh, Narendra Development University of Agriculture and Technology, Narendra Nagar, Kamarganj, Faizabad, 224001 Uttar Pradesh
Ranjit Singh, Himachal Pradesh Krishi Vishwa Vidyalaya, Palampur, 176062 Himachal Pradesh
Samar Singh, National Wastelands Development Board, Ministry of Environment and Forests, Paryavaran Bhavan, B Block-C.G.O. Complex, 110003 New Delhi
M.V.K. Sivamohan, Administrative Staff College of India, Bella Vista, Hyderabad, 500049 Andhra Pradesh
George Smith, Land and Water Engineering, International Crops Research Institute for the Semi-Arid Tropics (ICRISAT), Pantancheru PO, 504324 Andhra Pradesh
K.L. Srivastava, Farming Systems Research Program, International Crops Research Institute for the Semi-Arid Tropics (ICRISAT), Patancheru PO, 502324 Andhra Pradesh
S. Subramanian, Regional Research Station, Aruppukkottai, Tamil Nadu Agricultural University, Coimbatore, 641033 Tamil Nadu
S. Subramanian, Soil and Water Management Research Institute, Tamil Nadu Agricultural University, Kattuthottam, Thanjavur, 613501 Tamil Nadu
S. Subramanya, Watershed Development Programme, III Floor Podium Block, Visvesvarya Centre, Dr. Ambedkar Road, Bangalore, 560001 Karnataka
S.C. Trivedi, Himalayan Watershed Project, Government of Uttar Pradesh, Pauri, Uttar Pradesh

D.K. Uppal, Dr. Yashwant Singh Parmar University of Horticulture and Forestry, Solan, 173230 Himachal Pradesh
V.P.S. Verma, 151 Vasant Vihar-Indiranagar Forest Colony, Dehra Dun, 248006 Uttar Pradesh (germplasm and trials)

Indonesia

Amin Aziz, Pusat Pangembangan Agribisnis (PPA), Jalan Tebet Raya 57, Jakarta 12820
Chris Eijeumans, Vetiver Extension Programme, PPW/LTA-77 Development Project, Tromol Pos No. 3, Takengon 24514, Aceh Tengah (germplasm and trials)
Bambang Ismawan, Yayasan Bina Swadaya, PO Box 1456, Jakarta 10610
Pusat Pengembangan Kopi, Department of Agroresearch, c/o Section/Soils, The Coffee Development Centre of Aceh Tengah, Lta 77, Jr. Usman, Oekso/Tanah/Acru, Takengon 24582, Aceh Tengah (germplasm and trials)
Kedar N. Mutreja, ECI-Medan Urban Development Project, Jalan Babura Lama No. 8, Medan, North Sumatra (germplasm and trials)
Anton Soedjarwo, Yayasan Dian Desa, Jalan Kaliurang km 7, Jurugsari IV, Yogyakarta
Carol Stoney, Ford Foundation, PO Box 2030, Jakarta
Bambung Sukartiko, Office of Soil Conservation, Forestry Building, Manggala Wanabakti, Jalan Jend Gatot Subroto, Senayan, Jakarta Pusat
Dwiatmo Siswomartono, Directorate of Soil Conservation, Ministry of Forestry, Jalan Gatot Subroto, Jakarta 10270
Sidarto Wardoyo, Research Institute for Estate Crops, Taman Kencana 1, Bogor

Ireland

Gregory Dillon, 30 Collins Park, Abbeyfeale, County Limerick

Israel

Gerard N. Amzallag, Department of Botany and Applied Plant Physiology, Hebrew University, Guivat Ram 91904, Jerusalem 94551
Jiftah Ben-Asher, Centre for Desert Agrobiology, Ben Gurion University of Negev, PO Box 653, Beersheva
Ezra Henkin, Ministry of Agriculture, Soil Conservation and Drainage Division, Hakirya, Tel Aviv 61070

Italy

Volker Branscheid, Investment Center, Food and Agriculture Organization of the United Nations (FAO), Via delle Terme di Caracalla, 00100 Rome (germplasm)
R. Brinkman, Soil Resources, Management and Conservation Service, Food and Agriculture Organization of the United Nations (FAO), Via delle Terme di Caracalla, 00100 Rome
Geoffry Hawtin, International Board for Plant Genetic Resources, Food and Agriculture Organization of the United Nations (FAO), Via delle Sette Chiese 142, 00145 Rome
Gerhard Meier, Caritas Internationalis, Palazzo San Calisto, 120, Vatican City

Jamaica

Algernon V. Chin, Instituto Interamericano de Ciencias Agrícolas (IICA), PO Box 349, Kingston 6
Joseph R.R. Suah, Hillside Agriculture Project, Ministry of Agriculture, Hope Gardens, Kingston 6

Japan

Yuji Fujisawa, International Research Division, Research Council Secretariat, Ministry of Agriculture, Forestry and Fisheries, 1-2-1, Kasumigaseki, Chiyoda-Ku, Tokyo 100
Shiro Yoshida, Research and Development Division, Institute of International Cooperation, Japan International Cooperation Agency, International Cooperation Center, 10-5 Ichigaya Honmura-Cho, Shin Juku-Ku, Tokyo 162

Kenya

Rambaldi Giacomo, Lodagri Nairobi Branch, PO Box 67878, Nairobi (germplasm and trials)
Vernon Gibberd, E.M.I. Soil and Water Conservation Project, Ministry of Agriculture, Provincial Agricultural Headquarters, Eastern Province, PO Box 1199, Embu (germplasm and trials)
Enoch K. Kandie, Ministry of Agriculture, PO Box 30028, Nairobi
Simon Muchiru, African NGOs Environment Network (ANEN), PO Box 53844, Nairobi
C. Ndiritu, Kenya Agricultural Research Institute, PO Box 14733, Nairobi
Caleb O. Othieno, Tea Research Foundation of Kenya, PO Box 820, Kericho (germplasm and trials)
Hilda Munyua, Meka Rao, and Pedro Sanchez, International Centre for Research in Agroforestry (ICRAF), PO Box 30677, Nairobi (germplasm and trials)
Jeffrey R. Simpson, Australia-Funded Dryland Project, PO Box 41567, Nairobi

Kuwait

Ibrahim M. Hadi, Environment Protection Council, PO Box 24395, Safat

Laos

Jean Paul Boulanger, Micro Projects-Luang Prabang, PO Box 3705, Vientiane
Chanthaviphone Inthavong, Department of Forestry, PO Box 1034, Vientiane
Walter Roder, LAO-IRRI (International Rice Research Institute) Project, PO Box 600, Luang Prabang
John M. Schiller, Lao-IRRI Project, PO Box 4195, Vientiane (germplasm)

Lebanon

International Centre for Agricultural Research in the Dry Areas (ICARDA), Dalia Building, 2nd Floor, Rue Bashir El-Kassar, Beirut

Lesotho

Klaus Feldner, German Agency for Technical Cooperation (GTZ), PO Box 988, Maseru 100
Richard Holden, Soil and Water Conservation and Agroforestry Programme, PO Box 24, Maseru 100

Madagascar

Tom Bredero, c/o World Bank, BP 4140, Antananarivo (germplasm and trials)

Malawi

Malcolm J. Blackie, Agricultural Sciences, Rockefellar Foundation, PO Box 30721, Lilongwe 3 (germplasm and trials)
Stephen J. Carr, Christian Services Committee of Malawi, Private Bag 5, Zomba (germplasm and trials)
Francis W. M'Buka, World Bank Resident Mission, PO Box 30557, Capital City, Lilongwe (germplasm and trials)

Malaysia

K.F. Kon, CIBA-GEIGY Agricultural Experiment Station, Locked Bag, Rembau, Negri Sembican 71309 (germplasm and trials)
Azeez Abdul Ravoop, Institute for Advanced Studies, University of Malaya, Kuala Lumpur 59100 (germplasm and trials)
Zulkifli Haji Shamsuddin, Universiti Pertanian Malaysia, Serdang, Selangor 43400
Cheng Hai Teoh and Khairudin bin Hashim, Golden Hope Plantations Berhad, Prang Besar Research Station, Kajang, Selangor 43009 (germplasm and trials)
Cheriachangel Mathews, North Labis Estate, Labis, Johore 85300 (germplasm and trials)
Koh Tai Tong, Estate Department, Asia Oil Palm Sdn Bhd, Suite 120 Johor Tower, 15, Jln Geraja, Johor Bahru 80100 (germplasm and trials)
Ng Thai Tsiung, Agricultural Research Centre, Department of Agriculture, PO Box 977 Semongok, Kuching, Sarawak 93720 (germplasm and trials)
P.K. Yoon, Plant Science Division, Rubber Research Institute of Malaysia, Experiment Station, Sungei Buloh, Kuala Lumpur, Selangor 47000 (germplasm and trials)

Mali

West Africa Sorghum Improvement Program, International Crops Research Institute for the Semi-Arid Tropics (ICRISAT), BP 320, Bamako

Mauritius

Jean-Claude Autrey, Plant Pathology Division, Mauritius Sugar Industry Research Institute, Réduit (germplasm)
A. Kisto, National Federated Young Farmer's Club, Royal Road, Phoenix
T.L. Lamport, Essaims et Essences, Ltd., PO Box 47, Curepipe
Azad M. Osman, School of Agriculture, University of Mauritius, Réduit

Mexico

Salvador Gonzalez R., Collegon Trujillo #18, Atotomilco el alto, CP 47750, Jalisco
Arturo M. Moran, FIRCO, Zaragoza No. 58, Yurecuaro, Michoacan
Antonio Turrent Ferdinandez, Collegio del Valle, Insurgentes Sur 694-1000, Mexico 03100 D.F. (germplasm)
Johnathan Woolley, Centro Internacional de Mejoramiento de Maíz y Trigo (CIMMYT), PO Box 6-641, Mexico 06600 D.F.

Morocco

International Centre for Agricultural Research in the Dry Areas (ICARDA), BP 6299, Rabat Institute, Rabat

Nepal

Thomas Arens, World Neighbors, PO Box 916, Kathmandu (germplasm and trials)
Suresh Raj Chalise and Tej Partap, International Centre for Integrated Mountain Development (ICIMOD), GPO Box 3226, Kathmandu
R.N. Deo, Bhairawa Lumbini Groundwater Project, Amchalpur, Bhairawa, Lumbini Zone (germplasm and trials)
Benedikt Dolf, Helvetas, PO Box 113, Ekanta Kuna, Jawalakhel
Thakur Giri, Redd Barna-Nepal, PO Box 11, Tansen, Palpa (germplasm and trials)
Shesh Kanta Kafle, Farm Forestry Project, Institute of Forestry, PO Box 43, Kaski District, Gandaki Zone
Ram Mishra, Nigel Roberts, and Shambur Kumar Shrestha, World Bank Resident Mission, Kantipath, Kathmandu (germplasm and trials)
Gerold Muller, Bagmati Watershed Project-IDC, Ekanta Kuna, Jawalakhel, PO Box 730, Kathmandu (germplasm and trials)
Badri Nath Kayastha, No-Frills Development Consultants (NFDC), Manbhawan, Lalitpur, PO Box 3445, Kathmandu
S.B. Panday, Central Animal Nutrition Division, Khumaltar, Nepal (chemical analysis)
Krishna Sharma, Bhairawa Lumbini Groundwater Project, Amchalpur, Taulihawa, Lumbini Zone (germplasm and trials)
Shambhu Kumar Shrestha, Department of Agriculture, Harihar Bhavan, Pulchowk, Kathmandu (germplasm and trials)
Franz Thun, German Agency for Technical Cooperation (GTZ), PO Box 1457, Kathmandu
Saryug Pd. Yadav, Community Welfare and Development Society (CWDS), PO Box 5463, Kathmandu (germplasm and trials)

Netherlands

H.K. Jain, International Service for National Agricultural Research, PO Box 93375, The Hague 2509 AJ
Wim Spaan, Department of Land and Water Use, Wageningen Agriculture University, Nieuwe Kanaal II, Wageningen 6709 PA
Jan Diek van Mansvelt, Department of Ecological Agriculture, Haarweg 333, Wageningen 6709 RZ
L.J. van Veen, Internationaal Agrarisch Centrum, Lawickse Allee 11, PO Box 88, Wageningen 6700 AB (economics of vetiver oil)

Netherlands Antilles

Edward J.M.H. Berben, Department of Agriculture, Eilandgebied Bonaire, PO Box 43, Kralendijk, Bonaire

New Zealand

Grant B. Douglas, Department of Scientific and Industrial Research (DSIR), Private Bag, Palmerston, North
John C. Greenfield, 21 Reinga Road, Kerikeri, Northland (germplasm)
Donald E.K. Miller, Miller Environmental Consultants, 77 Shelley Street, Gisborne Zealand (germplasm and trials)

Niger

International Crops Research Institute for the Semi-Arid Tropics (ICRISAT) BP 12404, Niamey (hedge technology)

Nigeria

I. Okezie Akboundu, Division of Weed Science, International Institute of Tropical Agriculture (IITA), Oyo Road, PMB 5320, Ibadan
Kwesi Atta-Krah, Alley Farming Network for Tropical Africa (AFNETA), International Institute for Tropical Agriculture (IITA), Oyo Road, PMB 5320, Ibadan
Charles C. Ibe, Gully Erosion Afforestation Project, Experimental Research Station, PMB 7011, Umuahia, Abia State (germplasm and trials)
CIAT/IITA Cassava Program, c/o International Institute of Tropical Agriculture, PMB 5320, Oyo Road, Ibadan
T.E. Ekpenyong, Department of Animal Science, Alley Farming Network for Tropical Africa (AFNETA), University of Ibadan, Ibadan
Sister Elizabeth Fallon, Daughters of Charity St. Vincent De Paul, Reg. House, PO Box 123, Nchia, Eleme, Rivers State
Federal Agricultural Coordinating Unit, PMB 2277, Kanuna, Kaduna State (germplasm)
Jacob Ibrahim, Sokoto Agricultural Development Project, PMB 2245, Sokoto, Northwestern State
Osuagwu E. Iyke, Agro-Service Center, PO Box Umunama, Ezinihitte Mbaise, Imo
B. Kang, International Institute of Tropical Agriculture (IITA), Oyo Road, PMB 5320, Ibadan (germplasm and trials)
Lawal M. Marafa, Department of Forestry and Agricultural Land Resources, Department of Forestry, Wuse, Zone 1, PMB 135, Abuja, East Central State (germplasm and trials)
Bernard Nwadialo, Department of Social Science, University of Nigeria, Nsukka, Anambra State (germplasm and trials)
Soneye Alabi S. Okanlawon, Department of Geography and Planning, University of Lagos, Akoka-Yaba, Lagos (germplasm and trials)
Heanyl Okonkwo, Box 48 Oraifite, Nnewi LGA, Anambra State
Ladance B. Sunday, Ladson Farm, PO Box 954, Osogbo, Osun State (germplasm and trials)
Elisha D. Yakubu, Land Development Office, Kaduna Agricultural Development Project, PMB 1000, Garaje-Agban, Kagoro, Kaduna State (germplasm and trials)

Pakistan

Akhlaq Ali Khan, Hi-Tech Equipment (Private) Ltd., 3/49-Al-Yusuf Cham., Shahrah-Liaquat, Karachi
Jim Campbell, Barani—Agricultural Research and Development, National Agricultural Research Centre (NARC), Lab Wing, PO Box 1785, Islamabad
International Centre for Agricultural Research in the Dry Areas (ICARDA), Arid Zone Research Institute, Brewery Road, Box 362, Quetta

Panama

Ricardo Gavidia, Grounds Branch, Panama Canal Commission, Balboa

Papua New Guinea

Joe Yalgol Degemba, Simbu Agricultural Extension Support Project, PO Box 568, Kundiawa, Simbu Province
Friends of the Earth, PO Box 4028, Boroko
Anton Kaile, South Simbu Rural Development Project (SSRDP), PO Box 192, Kundiawa, Simbu Province
Ricky Kumung and Vaughan Redfern, Land Utilization Section, Department of Agriculture and Livestock, PO Box 1863, Boroko (germplasm and trials)

Mark Rosato Mandala, Foundation for the Peoples of the South Pacific, PO Box 297, Madang
Peter Metcalfe, c/o PNG Estates Ltd., PO Box 1131, Rabaul
John B. Mills, The Anisa Foundation, PO Box 26, Rabaul

Peru

Andean Bean Research Project, CIAT/IICA, Instituto Interamericano de Ciencias Agrícolas, AP 14-0185, Lima 14
Ernesto Diaz Falconi, Centro de Desarrollo Rural de Chincha, Ministry of Agriculture, Jiron Luis Massaro Gatnau 197, Chincha, Ica
INIAA/VITA/CIAT Pasture Research Program-Humid Tropics, Instituto Nacional de Investigación Agraria y Agroindustrial, AP 558, Pucallpa
Robert Rhoades, International Potato Center (CIP), Apartado 5969, Lima
Luis Miguel Saravia, Servicio de Apoyo de IRED-AL, Leon de la Fuente 110, Lima 17

Philippines

Ebert T. Bautista, Department of Environment and Natural Resources, Visayas Avenue, Diliman, Quezon City
Alex G. Coloma, Watershed Management and Erosion Control Project, NIA Campsite, Pantabangan, Nueva Ecija
Rogelio B. Dael, Upland Agriculture Site Management Unit, Central Visayas Regional Project Office, Bayawan, Negros Oriental (germplasm and trials)
John Dalton, ACIPHIL, 116 Legaspi Street, P&L Building 2nd Floor, Makati, Metro Manila
Alma Monica A. de la Paz, Kapwa Upliftment Foundation, Inc., Annex Building Room 2, Jacinto Campus, Ateneo de Davao University, Davao City (germplasm and trials)
Alois Goldberger, Abra Diocesan Rural Development, Bangued 2800, Abra (germplasm and trials)
Constancio D. de Guzman, Department of Horticulture, University of the Philippines at Los Baños, College, Laguna
Jose Galvez, NIA Watershed Act., National Irrigation Administration, ICC Building, Quezon City
Dennis P. Garrity, International Rice Research Institute (IRRI), PO Box 933, Manila 1099 (hedgerow technologies; germplasm and trials)
Alois Goldberger, Abra Diocesan Rural Development, Regional House Libbog, Bangued, Abra (germplasm and trials)
Alexander R. Madrigal, Department of Science and Technology—Region 4, Farcon Building, Rizal Avenue, San Pablo City, Laguna (germplasm and trials)
Leonard Q. Montemayor, Federation of Free Farmers, 41 Highland Drive, Blue Ridge, 1109 Quezon City
Enriqueta Perino, Ecosystems Research and Development Bureau, Department of Environment and Natural Resource (DEHR), College, Laguna
Robert John Sims, Misamis Oriental State College of Agriculture and Technology (MOSCAT), Claveria, Misamis Oriental 9004 (germplasm and trials)
Jose Tabago, Department of Agricultural Engineering, Central Luzon State University, Muñoz, Nueva Ecija (germplasm and trials)
Paul P.S. Teng, International Rice Research Institute (IRRI), PO Box 933, Manila 1099 (pathology)
Ly Tung, Farm and Resource Management Institute (FARMI), Visayas State College of Agriculture (VisCA), Baybay, Leyte 6521-A (germplasm and trials)
John L. Waggaman, Barrio San Pedro, Bauan, Batangas 4201, (germplasm and trials)
Terence Woodhead, Department of Soil Physics, University of the Philippines at Los Baños, College, Laguna

Portugal

Graham Quinn, Sitio Atlas 179, Garajau, 9125 Canico, Madeira

Rwanda

Institut de recherche et de éducation et développement (IRED), BP 257, Cyangugu
Francois Ndolimana, Division of Fertilizers, Ministry of Agriculture, BP 1648, Kigali
Vincent Nyamulinda, c/o OPYRWA, PO Box 79, Ruhengeri (germplasm and trials)

Saudi Arabia

John Bowlin, World Bank Field Office, PO Box 5900, Riyadh 11432

Scotland

Landwise Scotland, Brig O Lead, Forbes, Alford AB3 8PD
Keith A. Smith, The Edinburgh School of Agriculture, West Mains Road, Edinburgh EH9 3JG

Senegal

Thierno Kane, Federation de associations du Fouta le développement, BP 3865, Dakar
Amadou Seck, Rodale International, BP 237, Thies

Sierra Leone

Kiets Hall, FAO Agroforestry Project, c/o United Nations Development Programme (UNDP), PO Box 1011, Freetown
Vijay Wijegoonaratna, ILO Rural Project, c/o UNDP, PO Box 1011, Freetown

Singapore

Patrick Y. Durand, Southeast Asia Representative Office, Centre Technique Forestier Tropical (CTFT), #14-275 Selegie Com., 257 Selegie Road

Solomon Islands

Bill Cogger, Principal, National Agricultural Training Institute, Auki, Malaita
Morgan Wairiu, Dodo Creek Research Station, Ministry of Agriculture and Lands, PO Box G 13, Honiara

South Africa

C.W. Browne, Development Aid, PO Box 384, Pretoria 0001
Daya Chetty, House of Delegates, Private Bag X0580, Umzinto 4200
S. Christie, Council for Scientific and Industrial Research (CSIR), Private Bag X520, Sabie 1260
John Erskine, University of Natal, PO Box 375, Pietermaritzburg 3200
Chris Nicolson, PO Box 11015, Dorspruit 3206

C. Ntsane, QwaQwa Government, Private Bag X816, Witsishoek 9870
Noel Oettle, University of Natal, PO Box 375, Pietermaritzburg 3200
A.J. Pembroke, Botanic Gardens, 70 St. Thomas Road, Durban 4001
Zoomie Robert, PO Box 405, Umhlali 4390
Abbas Shaker, Private Bag X5002, Umtata, Transkei
Rudi Snyman, Agricor, Private Bag X2137, Mmabatho 8686, Bophuthatswana
Mr. Swanepoel, Ministry of Water and Forestry, Private Bag X9052, Cape Town 8000
Anthony Tantum, Vetiver Grass Stabilization cc, PO Box 167, Howick 3290 (germplasm and trials)
P. van Eldik, Board of Control, Potchefstroom University, Private Bag X6001, Potchefstroom
J.J.P. Van Wyk, Research Institute for Reclamation Ecology, Potchefstroom University, Potchefstroom 2520 (germplasm; Eragrostis)
W.P.J. Wessels, Department of Soil Science, University of Stellenbosch, Stellenbosch 7600

South Korea

Wong Kwon Yong, Department of Agronomy, Seoul National University, Suwon 440-744

Sri Lanka

M.B. Adikaram, Nation Builders Association, 48 Hill Street, Kandy
Godfrey Gunatilaka, Marga Institute, 61 Isipatana Mawatha, Colombo 5
Roberto Lenton, International Irrigation Management Institute (IIMI), Digana Village, Digana via Kandy
Lee Moncaster, CARE-Sri Lanka, Vilasitha Nivasa 2nd Floor, 375 Havelock Road, Colombo 6
K. Rajapakse, 147/8 Dharmsoka Mawatha, Lewella, Kandy
L.M. Samarasindhe, NGO Council of Sri Lanka, 380 Bauddhaloka Mawatha, Colombo 7

St. Lucia

Harry Atchinson, c/o St. Lucia Banana Association, Castries
Gabriel Charles, Tropical Forest Action Programme, Ministry of Agriculture, PO Box 1537, Castries
Barton Clarke, Caribbean Agricultural Research and Development Institute (CARDI), Windban Research Station, PO Box 971, Castries
Standley Mullings, Stanthur and Company, Limited, Columbus Square, Castries

St. Vincent

Lennox Diasley, Ministry of Trade, Industry, and Agriculture, Kingstown
Leonard Jack, Caribbean Development Foundation, PO Box 920, Kingstown
Conrad Simon, Ministry of Trade, Industry, and Agriculture, Kingstown

Sudan

Ceasar Kenyi Draku, Extension Department, Ministry of Agriculture, PO Box 194, Kassala, Eastern Sudan (germplasm and trials)
Omer Elgoni, Range and Pasture Administration, PO Box 2513, Khartoum

K.D. Shepherd, Jebal Marra Rural Development Project, PO Box 9010 (K.T.I.), Khartoum
Abdalla Sulliman Elawad, Islamic African Relief Agency, PO Box 3372, Khartoum

Sweden

Karina Francis, Agriculture Division, Swedish International Development Authority (SIDA), Birger Jarlsgatan 61, Stockholm S-105 25

Switzerland

Christopher Gibbs, Aga Khan Foundation, PO Box 435, Geneva 1211
Robert Quinlan, International Office, Catholic Relief Services, 11 rue Cornavin, Geneva 1201
Cyril Ritchie, Esperanza Duran, International Council of Voluntary Agencies (ICVA), 13 rue Gautier, Geneva 1201
Andreas Zoschke, CIBA-GEIGY Agricultural Division-Plant Protection, Basle 4002

Syria

Nasrat Fadda, International Centre for Agricultural Research in the Dry Areas (ICARDA), PO Box 5466, Aleppo

Taiwan

C.H. Huang, Food and Fertilizer Technology Center, 14 Wenchow Street, 5th Floor, Taipei

Tanzania

Ted Angen, Selian Agricultural Research Center, Box 6160, Arusha (germplasm and trials)
Janet Cundall, Cashew Improvement Project, PO Box 608, Mtwara
Paul Richardt Jensen, Hima-Danida-iringa Soil and Water Conservation Project, PO Box 1187, Iringa (germplasm and trials)
L. Nshubemuki, Tanzania Forestry Research Institute, PO Box 45, Maftinga

Thailand

Christoph Backhaus, Thai-German Highland Development Programme, PO Box 67, Chiang Mai 50000
Biomass Users Network, Regional Office for Asia, PO Box 275, Bangkok
Arthorn Boonsaner, Watershed Division, Royal Forestry Department, Phaholyothin Road, Bangkok 10900
Narong Chomchalow, Regional Office for Asia and the Pacific (RAPA), Food and Agriculture Organization of the United Nations (FAO), Maliwan Mansion, Phra Atit Road, Bangkok 10200 (germplasm)
Tom Drahman, CARE-Thailand, 246/4 Soi, Rama 6 Road, Bangkok 10400
Bhakdi Lusanandana, Agricultural Land Reform Office, Thanon Katchadamnoen Nok, Bangkok 10200
Malee Nanakorn, Department of Botany, Kasetsart University, Bangkhen, Bangkok 10900
Apichai Thirathon, Office of Land Development Region 9, Rim Mae Nam Road, Nakornsawan Tok, A. Muang 60000, Nakornsawan (germplasm and trials)

Togo

Boukari Ayessaki, RAFIA, BP 43, Dapaong (germplasm and trials)
Tchemi Wouro, Ministre du Developpement Rural, Lome

Trinidad and Tobago

T. Fergerson, Department of Crop Science, University of the West Indies, St. Augustine
Joseph I. Lindsay, Department of Soil Science, University of the West Indies, St. Augustine
Selwyn Dardiane, Director of Forests, Trinidad and Tobago, Port of Spain

Turkey

Faik Ahmet Ayaz, Department of Biology, Faculty of Sciences, Karadeniz Technical University, 61080 Trabzon
Nazmi Demir, Ministry of Agriculture, Forestry and Rural Affairs, Ankara
J. Sahin, TUBITAK-DEBCAG, Istanbul Cad. No. 88, Ankara

Uganda

Ruth Mubiru, Uganda Women Tree Planting Movement, PO Box 10351, Kampala
J. Mugwera, Faculty of Agriculture, Makerere University, PO Box 7082, Kampala

United States

E.E. Alberts, Cropping Systems and Water Quality Research, ARS, USDA, University of Missouri, Columbia, Missouri 65211 (quantitative measure of runoff and erosion)
Aledra Allen, Agriculture Division (ASTAG), Asia-Technical Department, The World Bank, 1818 H Street, N.W., Washington, D.C. 20433 (Vetiver Network)
Horace Austin, Soil Conservation Service, U.S. Department of Agriculture (USDA), 3737 Government Street, Alexandria, Louisiana 71302
Deepak Bhatnagar, Food and Feed Safety Research, Agricultural Research Service (ARS), U.S. Department of Agriculture (USDA), PO Box 19687, Room 2129, New Orleans, Louisana 70179
B.B. Billingsley, Coffeeville Plant Material Center, SCS, USDA, Route 3, Box 215A, Coffeeville, Mississippi 38922
Al Binger, Biomass Users Network, PO Box 33308, Washington, D.C. 20033
Eugene Le Blanc, 1427 Huey 75, Sunshine, Louisiana 70780 (Sunshine germplasm)
Mark Le Blanc, Department of Horticulture, Louisiana State University, Baton Rouge, Louisiana 70803 (germplasm and trials)
Geric Boucard, Texarome Inc., Leakey, Texas 78873 (mechanized propagation, planting, and harvesting; agronomy; oil distillation; germplasm)
John Boutwell, Research and Laboratory Services Division, U.S. Bureau of Reclamation, PO Box 25007, Denver, Colorado 80225-0007 (germplasm)
Pat Boyd, National Plant Germplasm Quarantine Laboratory, Plant Science Institute, ARS, USDA 11601 Old Pond Drive, Glenn Dale, Maryland 20769-9157
James L. Brewbaker, Nitrogen Fixing Tree Association, PO Box 680, Waimanalo, Hawaii 96795
Lâle Aka Burk, Department of Chemistry, Smith College, Northampton, Massachusetts 01063 (vetiver oil chemistry)
Lewis Campbell, Rural Engineering, West African Agriculture Division, World Bank, 1818 H Street, N.W., Washington, D.C. 20433 (vetiver in Caribbean)
Thomas M. Catterson, Forestry/Natural Resources, Associates in Rural Development, Inc. (ARD), RD #1, Box 44, Clinton, New York 13323, USA (Haiti)
Carol Cox, Ecology Action, 5798 Ridgewood Road, Willits, California 95490 (germplasm)

Seth M. Dabney, USDA Sedimentation Laboratory, ARS, USDA, PO Box 1157, Oxford, Mississippi 38655 (germplasm and trials)
Gerrit Davidse, Missouri Botanical Garden, PO Box 299, St. Louis, Missouri 63166-0299 (taxonomy; vetiver in Central and North America)
Andre Delgado, Office of International Cooperation and Development (OICD), U.S. Department of Agriculture, Room 220, McGregor Building, Washington, D.C. 20250-4300
Kittie S. Derstine, Golden Meadows Plant Materials Center, SCS, USDA, PO Box 2202, Galliano, Louisiana 70354 (germplasm and growth studies)
James DeVries, Heifer Project International, PO Box 808, Little Rock, Arkansas 72203
Mel Duvall, Department of Biology, George Mason University, Fairfax, Virginia 22030 (agrostology)
James Eagan, 26411 Robin Street, Esparto, California 95627 (germplasm)
Thomas Eisner, Division of Biological Sciences, Seeley G. Mudd Building, Cornell University, Ithaca, New York 14853 (chemistry of vetiver oil)
E.D.M. Fletcher and John Muntz, SIFAT, Lineville, Alabama 36266 (germplasm and trials)
Connie Fitz, Box 505, Woodstock, Vermont 05091 (vetiver history)
Donald Fryrear, Big Spring Experiment Station, ARS, USDA, Box 909, Big Spring, Texas 79721 (wind erosion)
Levi Glover, SCS, USDA, Ft. Valley State College, PO Box 4061, Ft. Valley, Georgia 31030
Richard G. Grimshaw, Agriculture Division (ASTAG), Asia—Technical Department, The World Bank, 1818 H Street, N.W., Washington, D.C. 20433 (Vetiver Network Coordinator)
Roger Hanson, Tropsoils Program, University of North Carolina, Box 7619, Raleigh, North Carolina 27695
Richard R. Harwood, Department of Crop and Soil Sciences, Plant and Soil Science Building, Michigan State University, East Lansing, Michigan 48824-1325 (agricultural systems; germplasm)
Patrick Emeka Igbokwe, Department of Agriculture, Alcorn State University, PO Box 625, Lorman, Mississippi 39096 (germplasm and trials)
John Jeavons, Ecology Action, 5798 Ridgewood Road, Willits, California 95490
Loyd Johnson, Route 3, Box 486, Somerville, Alabama 35670
Robert J. Joy, SCS, USDA, Box 236, Hoolehua, Molokai, Hawaii 96729 (germplasm and trials)
W. Doral Kemper, Soil and Water Conservation Program, Plant and Natural Resource Sciences, ARS, USDA, Beltsville Agricultural Research Center, Beltsville, Maryland 20705 (vegetative hedges)
Larry Kramer, Cropping Systems and Water Quality Research, ARS, USDA, University of Missouri, Columbia, Missouri 65211 (grass hedges in temperate agriculture)
Stephen Kresovich, Plant Genetic Resource Unit, ARS, USDA, New York State Agricultural Research Station, Cornell University, Geneva, New York 14456-0462 (DNA analysis of genetic variation; clonal identification)
J.H. Krishna, Water Research Center, University of the Virgin Islands, St. Thomas, Virgin Islands
David L. Leonard, LUPE Project—Honduras, Department 236, PO Box 025320, Miami, Florida 33136 (germplasm and trials)
Gilbert Lovell, Southern Regional Plant Introduction Station, ARS, USDA, 1109 Experiment Street, Griffin, Georgia 30223-1797 (germplasm0
William B. Magrath, Environmental Policy Research Division, The World Bank -S-3065, 1818 H Street, NW, Washington, D.C. 20433 (economics)
Dan Mason, Wetland Research Project, PO Box 225, Wadsworth, Illinois 60083 (vetiver wetlands ecology)
Mike Materne, SCS, USDA, PO Box 16030 University Station, Baton Rouge, Louisiana 70803 (germplasm and trials)
John Mayernak, New Mexico State University, HCR 30 Box 61, Tucumcari, New Mexico 88401 (germplasm and trials)

Charles B. McCants, 201 Merwin Road, Raleigh, North Carolina 27606

Colin McClung, Winrock International, 1611 N. Kent Street, Arlington, Virginia 22209

Richard N. Middleton, Kalbermatten Associates Inc., 2327 Pondside Terrace, Silver Spring, Maryland 20906 (urban flood mitigation and erosion control)

Keith McGregor, USDA Sedimentation Laboratory, ARS, USDA, PO Box 1157, Oxford, Mississippi 38655 (quantitative measure of runoff and erosion)

Don Meyer, USDA Sedimentation Laboratory, ARS, USDA, PO Box 1157, Oxford, Mississippi 38655 (hedge hydrolics)

Seiichi Miyamoto, Texas Agricultural Experiment Station, 1380 A&M Road, El Paso, Texas 79927 (physiology; salt and mineral tolerances; germplasm)

R. Muniappan, Agricultural Experiment Station, University of Guam, Mangilao, Guam 96913 (germplasm and trials)

Mary Musgrave, Department of Plant Pathology and Physiology, Life Sciences Building, Louisiana State University, Baton Rouge, Louisiana 70803 (stress physiology and water-logging)

Charles Owsley, Americus Plant Materials Center, SCS, USDA, Route 6, Box 417, Americus, Georgia 31709

James H. Perkins, South Texas Plant Material Center, SCS, USDA, Texas A&I University, Kingsville, Texas 78363 (germplasm and trials)

J.S. Peterson, National Plant Materials Center, SCS, USDA, Building 509, BARC East, Beltsville, Maryland 20705 (germplasm; botany)

Kathleen Plavcan, Pfizer Corporation, 630 Flushing Avenue, Brooklyn, New York 11206 (chemisty)

Hugh Popenoe, International Program in Agriculture, 3028 McCarty Hall, University of Florida, Gainesville, Florida 32611 (germplasm and trials)

Martin L. Price, Educational Concerns for Hunger Organization (ECHO), 17430 Durrance Road, North Fort Myers, Florida 33917 (technical advice and locating germplasm)

Errol G. Rhoden, 308 Milbank Hall, Tuskegee University, Tuskegee, Alabama 36088 (germplasm and trials)

Jerry C. Ritchie, Hydrology Laboratory, ARS, USDA, Building 265 Room 205, BARC East, Beltsville Agricultural Research Station, Beltsville, Maryland 20705

Herbert Ross, SCS, USDA, Alabama A&M University, PO Box 183, Normal, Alabama 35762

John M. Safley, Ecological Sciences, SCS, USDA, PO Box 2890, Washington, D.C. 20013

J. Eric Scherer, National Plant Materials Center, SCS, USDA, Building 509, BARC East, Beltsville, Maryland 20705 (germplasm)

David Schumann, Forest Products Laboratory, U.S. Forest Service, U.S. Department of Agriculture, One Gifford Pinchot Drive, Madison, Wisconsin 53705-2398

W. Curtis Sharp, National Plant Materials Laboratory, SCS, USDA, PO Box 2890, Washington, D.C. 20013

Holly Shimizu, U.S. Botanic Gardens, First and Canal Street, SW, Washington, D.C. 20024 (germplasm)

John Silvius, Cederville College, Cederville, Ohio 45314

Jackie L. Smith, Directorate of Engineering and Housing, ATTN: AFZX-DE-E, U.S. Army, Fort Polk, Louisiana 71459-7100

James Smyle, Agriculture Division (ASTAG), Asia—Technical Department, The World Bank, 1818 H Street, N.W., Washington, D.C. 20433 (Vetiver Network)

Edward D. Surrency, SCS, USDA, 355 E. Hancock Avenue, Box 13, Athens, Georgia 30601

Prabmakar Tamboli, East Asia Pacific Agricultural Operations (EA3AG), World Bank, 1818 H Street, N.W., Washington, D.C. 20433 (germplasm and trials)

Grant Thomas, Department of Agronomy, University of Kentucky, Lexington, Kentucky 40506 (vegetative barriers)

Suresh C. Tiwari, Department of Agriculture, Alcorn State University, Lorman, Mississippi 39096

B. Dean Treadwell, Sove Te Haiti, PO Box 407103, Ft. Lauderdale, Florida 33340 (germplasm and trials)

Arnold Tschanz, Animal and Plant Health Inspection Service, USDA, Federal Building Room 625, 6505 Bellcrest Road, Hyattsville, Maryland 20782
Goro Uehara, Department of Agronomy and Soil Science, 1910 East-West Road, University of Hawaii, Honolulu, Hawaii 96822
Remko Vonk, Agriculture and Natural Resources, CARE, 660 First Avenue, New York, New York 10016
Howard Waterworth, National Plant Germplasm Quarantine Laboratory, ARS, USDA, 11601 Old Pond Drive, Glenn Dale, Maryland 20769
James A. Wolfe, SCS, USDA, Federal Building Suite 1321, 100 West Capitol Street, Jackson, Mississippi 39269 (germplasm and trials)

Vietnam

Dang Thanh Binh, Office Manager of Rural Development, Department of Agricultural Development, State Bank of Vietnam, Hanoi
William John Leith, Plantation and Soil Conservation Project, Vietnam-Sweden Forestry Cooperation Programme, c/o Interforest, Bai Bang, PO Box 1226, Nana P.O., Bangkok 10112, Thailand (germplasm)
Bui Quang Toan, NIAPP Office, 6 Nguyen cong Tru, Hanoi

Western Samoa

Tupuala Tavita, Department of Agriculture, Forests and Fisheries, PO Box 3017, Apia

Yemen

Mohammed Surmi, PO Box 3816, Sana

Zambia

Glenn Allison, District Development Support Program, PO Box 450148, Mpika (germplasm and trials)
George Richard Gray, Masstock (Zambia) Ltd., PO Box 34756, Lusaka (germplasm and trials)
Msanfu Research Station, Msanfu (germplasm)

Zimbabwe

David Pell Goodwin, Department of Surveying, University of Zimbabwe, PO Box MP 167, Mount Pleasant, Harare (germplasm and trials)
Jano L. Labat, Sugar Cane Farm, Chiware Holdings Put Ltd., PO Box 14, Chiredzi (germplasm and trials)
A.R. Maclaurin, Faculty of Agriculture, Department of Crop Science, University of Zimbabwe, PO Box MP 167, Mount Pleasant, Harare
Matopos Research Station, Department of Research and Special Services, PB K 5137, Bulawayo (germplasm)
Anthony O'Brien, Henderson Research Station, Ministry of Agriculture, PB 2004, Mazowe (germplasm)
David Scott, Chipinga Farm, Chipinga (germplasm and trials)
H. Vogel, Conservation Tillage Sustainable Crop Production Service-German Agency for Technical Cooperation (GTZ), PO Box 415, Borrowdale, Harare
John Wilson, Fambidzanai Training Centre, PO Box 8515, Causeway, Harare (germplasm and trials)

Appendix E
Biographical Sketches

NORMAN BORLAUG (*Chairman*) is a consultant in the Office of the Director General, Centro Internacional de Mejoramiento de Maíz y Trigo (CIMMYT) in Mexico City and a professor at Texas A&M University. A specialist in wheat breeding, agronomy, plant pathology, and other areas, he is one of the best-known spokesmen and ambassadors for tropical agriculture. He is particularly renowned for creating the high-yielding wheat varieties that have transformed the grain supplies of India, Pakistan, and other nations. A native of Cresco, Iowa, he is now a citizen of both the United States and Mexico and is the recipient of more than 30 honorary degrees. In 1970 he was awarded the Nobel Prize for Peace.

RATTAN LAL has been a member of the department of agronomy at Ohio State University since 1987. In 1968 he received his Ph.D. in agronomy (soil physics) from Ohio State University. From 1969 to 1987, he worked as a soil physicist and coordinator of upland production systems with the International Institute for Tropical Agriculture (IITA) in Ibadan, Nigeria. His research interests include soil erosion and its control, soil structure and management, soil compaction and drainage, ecological impact of tropical deforestation, viable alternatives to shifting cultivation, and sustainable management of soil and water resources.

DAVID PIMENTEL is professor of insect ecology and agricultural sciences at Cornell University, Ithaca, New York. He received his Ph.D. in entomology from Cornell in 1951 and was chief of the U.S. Public Health Service (USPHS) tropical research laboratory in San Juan, Puerto Rico, and project leader of the USPHS technology development laboratory in Savannah, Georgia, before joining the department of entomology and limnology at Cornell. His particular interests are ecosystems management and pollution control, energy and land resources in the food system, and pest control. He has served as both chairman and member of numerous panels and committees of the National Research Council, including some on biology and renewable resources, agriculture and the environment, and innovative mosquito control. He served as chairman of the Board on Science and Technology for International Development from 1975-1980 and of the Environmental Studies Board of the National Research Council from 1981-1982.

HUGH POPENOE is professor of soils, agronomy, botany, and geography, and director of the Center for Tropical Agriculture and International Programs (Agriculture) at the University of Florida. He received his Ph.D. in soils science from the University of Florida in 1960. Since then his principal research interest has been in the area of tropical agriculture and land use. His early work on shifting cultivation is one of the major contributions to this system. He has traveled and worked in most of the countries in the tropical areas of Latin America, Asia, and Africa. His current interests include improving indigenous agricultural systems of small landholders, particularly with the integration of livestock and crops. Currently, he is on the international advisory committee of the National Science Foundation and serves as U.S. board member for the International Foundation of Science.

NOEL D. VIETMEYER, study director and technical writer for this study, is a senior program officer of the Board on Science and Technology for International Development. A New Zealander with a Ph.D. in organic chemistry from the University of California, Berkeley, he now works on innovations in science and technology that are important for the future of developing countries.

The BOSTID Innovation Program

Since its inception in 1970, BOSTID has had a small project to evaluate innovations that could help the Third World. Formerly known as the Advisory Committee on Technology Innovation (ACTI), this small program has been identifying unconventional developments in science and technology that might help solve specific developing-country problems. In a sense, it acts as an "innovation scout"—providing information on options that should be tested or incorporated into activities in Africa, Asia, and Latin America.

So far, the BOSTID innovation program has published about 40 reports, covering, among other things, underexploited crops, trees, and animal resources, as well as energy production and use. Each book is produced by a committee of scientists and technologists (including both skeptics and proponents), with scores (often hundreds) of researchers contributing their knowledge and recommendations through correspondence and meetings.

These reports are aimed at providing reliable and balanced information, much of it not readily available elsewhere and some of it never before recorded. In its two decades of existence, this program has distributed approximately 350,000 copies of its reports. Among other things, it has introduced to the world grossly neglected plant species such as jojoba, guayule, leucaena, mangium, amaranth, and the winged bean.

BOSTID's innovation books, although often quite detailed, are designed to be easy to read and understand. They are produced in an attractive, eye-catching format, their text and language carefully crafted to reach a readership that is uninitiated in the given field. In addition, most are illustrated in a way that helps readers deduce their message from the pictures and captions, and most have brief, carefully selected bibliographies, as well as lists of research contacts that lead readers to further information.

By and large, these books aim to catalyze actions within the Third World, but they usually are useful in the United States, Europe, Japan, and other industrialized nations as well.

To date, the BOSTID innovation project on underexploited Third World resources (Noel Vietmeyer, director and scientific editor) has produced the following reports.

Ferrocement: Applications in Developing Countries (1973). 104 pp.
Mosquito Control: Perspectives for Developing Countries (1973). 76 pp.
Some Prospects for Aquatic Weed Management in Guyana (1974). 52 pp.

Roofing in Developing Countries: Research for New Technologies (1974). 84 pp.
An International Centre for Manatee Research (1974). 38 pp.
More Water for Arid Lands (1974). 165 pp.
Products from Jojoba (1975). 38 pp.
Underexploited Tropical Plants (1975). 199 pp.
The Winged Bean (1975). 51 pp.
Natural Products for Sri Lanka's Future (1975). 53 pp.
Making Aquatic Weeds Useful (1976). 183 pp.
Guayule: An Alternative Source of Natural Rubber (1976). 92 pp.
Aquatic Weed Management: Some Prospects for the Sudan (1976). 57 pp.
Ferrocement: A Versatile Construction Material (1976). 106 pp.
More Water for Arid Lands (French edition, 1977). 164 pp.
Leucaena: Promising Forage and Tree Crop for the Tropics (1977). 123 pp.
Natural Products for Trinidad and the Caribbean (1979). 50 pp.
Tropical Legumes (1979). 342 pp.
Firewood Crops: Shrub and Tree Species for Energy Production (volume 1, 1980). 249 pp.
Water Buffalo: New Prospects for an Underutilized Animal (1981). 126 pp.
Sowing Forests from the Air (1981). 71 pp.
Producer Gas: Another Fuel for Motor Transport (1983). 109 pp.
Producer Gas Bibliography (1983). 50 pp.
The Winged Bean: A High-Protein Crop for the Humid Tropics (1981). 58 pp.
Mangium and Other Fast-Growing Acacias (1983). 72 pp.
Calliandra: A Versatile Tree for the Humid Tropics (1983). 60 pp.
Butterfly Farming in Papua New Guinea (1983). 42 pp.
Crocodiles as a Resource for the Tropics (1983). 69 pp.
Little-Known Asian Animals With Promising Economic Future (1983). 145 pp.
Casuarinas: Nitrogen-Fixing Trees for Adverse Sites (1983). 128 pp.
Amaranth: Modern Prospects for an Ancient Crop (1983). 90 pp.
Leucaena: Promising Forage and Tree Crop (Second edition, 1984). 110 pp.
Jojoba: A New Crop for Arid Lands (1985). 112 pp.
Quality-Protein Maize (1988). 112 pp.
Triticale: A Promising Addition to the World's Cereal Grains (1989). 113 pp.
Lost Crops of the Incas: Little-Known Plants of the Andes with Promise for Worldwide Cultivation (1989). 427 pp.

Microlivestock: Little-Known Small Animals with a Promising Economic Future (1991). 468 pp.
Neem: A Tree for Solving Global Problems (1992). 151 pp.
Vetiver: A Thin Green Line Against Erosion (1992).
Lost Crops of Africa: Volume 1—Grains (1992).
Lost Crops of Africa: Volume 2—Cultivated Fruits (1992).
Lost Crops of Africa: Volume 3—Wild Fruits (1992).
Lost Crops of Africa: Volume 4—Vegetables (In preparation)
Lost Crops of Africa: Volume 5—Legumes (In preparation)
Lost Crops of Africa: Volume 6—Roots and Tubers (In preparation)
Foods of the Future: Tropical Fruits (In preparation)
Foods of the Future: Tropical Fruits to Help the World (In preparation)

Board on Science and Technology for International Development

ALEXANDER SHAKOW, Director, External Affairs, The World Bank, Washington, D.C., *Chairman*

Members

PATRICIA BARNES-MCCONNELL, Director, Bean/Cowpea CRSP, Michigan State University, East Lansing, Michigan

JORDAN J. BARUCH, President, Jordan Baruch Associates, Washington, D.C.

BARRY BLOOM, Professor, Department of Microbiology, Albert Einstein College of Medicine, Bronx, New York

JANE BORTNICK, Assistant Chief, Congressional Research Service, Library of Congress, Washington, D.C.

GEORGE T. CURLIN, National Institutes of Allergy and Infectious Diseases, National Institutes of Health, Bethesda, Maryland

DIRK FRANKENBERG, Director, Marine Science Program, University of North Carolina at Chapel Hill, Chapel Hill, North Carolina

RALPH HARDY, President, Boyce-Thompson Institute for Plant Research, Inc., Ithaca, New York

FREDERICK HORNE, Dean, College of Sciences, Oregon State University, Corvallis, Oregon

ELLEN MESSER, Allan Shaw Feinstein World Hunger Program, Brown University, Providence, Rhode Island

CHARLES C. MUSCOPLAT, Executive Vice President, MCI Pharma, Inc., Minneapolis, Minnesota

JAMES QUINN, Amos Tuck School of Business, Dartmouth College, Hanover, New Hampshire

VERNON RUTTAN, Regents Professor, Department of Agriculture and Applied Economics, University of Minnesota, Saint Paul, Minnesota

ANTHONY SAN PIETRO, Professor of Plant Biochemistry, Department of Biology, Indiana University, Bloomington, Indiana

ERNEST SMERDON, College of Engineering and Mines, University of Arizona, Tucson, Arizona

GERALD P. DINEEN, Foreign Secretary, National Academy of Engineering, Washington, D.C., *ex officio*

JAMES WYNGAARDEN, Chairman, Office of International Affairs, National Academy of Sciences, National Research Council, Washington, D.C., *ex officio*

BOSTID Publications and Information Services (FO-2060Z)
National Research Council
2101 Constitution Avenue, N.W.
Washington, D.C. 20418 USA

How to Order BOSTID Reports

BOSTID manages programs with developing countries on behalf of the U.S. National Research Council. Reports published by BOSTID are sponsored in most instances by the U.S. Agency for International Development. They are intended for distribution to readers in developing countries who are affiliated with governmental, educational, or research institutions, and who have professional interest in the subject areas treated by the reports.

BOSTID books are available from selected international distributors. For more efficient and expedient service, please place your order with your local distributor. Requestors from areas not yet represented by a distributor should send their orders directly to BOSTID at the above address.

Energy

33. **Alcohol Fuels: Options for Developing Countries.** 1983, 128 pp. Examines the potential for the production and utilization of alcohol fuels in developing countries. Includes information on various tropical crops and their conversion to alcohols through both traditional and novel processes. ISBN 0-309-04160-0.

36. **Producer Gas: Another Fuel for Motor Transport.** 1983, 112 pp. During World War II Europe and Asia used wood, charcoal, and coal to fuel over a million gasoline and diesel vehicles. However, the technology has since been virtually forgotten. This report reviews producer gas and its modern potential. ISBN 0-309-04161-9.

56. **The Diffusion of Biomass Energy Technologies in Developing Countries.** 1984, 120 pp. Examines economic, cultural, and political factors that affect the introduction of biomass-based energy technologies in developing countries. It includes information on the opportunities for these technologies as well as conclusions and recommendations for their application. ISBN 0-309-04253-4.

Technology Options

14. **More Water for Arid Lands: Promising Technologies and Research Opportunities.** 1974, 153 pp. Outlines little-known but promising technologies to supply and conserve water in arid areas. ISBN 0-309-04151-1.

21. **Making Aquatic Weeds Useful: Some Perspectives for Developing Countries.** 1976, 175 pp. Describes ways to exploit aquatic weeds for grazing and by harvesting and processing for use as compost, animal feed, pulp, paper, and fuel. Also describes utilization for sewage and industrial wastewater. ISBN 0-309-04153-X.

34. **Priorities in Biotechnology Research for International Development: Proceedings of a Workshop.** 1982, 261 pp. Report of a workshop organized to examine opportunities for biotechnology research in six areas: 1) vaccines, 2) animal production, 3) monoclonal antibodies, 4) energy, 5) biological nitrogen fixation, and 6) plant cell and tissue culture. ISBN 0-309-04256-9.

61. **Fisheries Technologies for Developing Countries.** 1987, 167 pp. Identifies newer technologies in boat building, fishing gear and methods, coastal mariculture, artificial reefs and fish aggregating devices, and processing and preservation of the catch. The emphasis is on practices suitable for artisanal fisheries. ISBN 0-309-04260-7.

73. **Applications of Biotechnology to Traditional Fermented Foods.** 1992, 207 pp. Microbial fermentations have been used to produce or preserve foods and beverages for thousands of years. New techniques in biotechnology allow better understanding of these transformations so that safer, more nutritious products can be obtained. This report examines new developments in traditional fermented foods. ISBN 0-309-04685-8.

Plants

47. **Amaranth: Modern Prospects for an Ancient Crop.** 1983, 81 pp. Before the time of Cortez, grain amaranths were staple foods of the Aztec and Inca. Today this nutritious food has a bright future. The report discusses vegetable amaranths also. ISBN 0-309-04171-6.

53. **Jojoba: New Crop for Arid Lands.** 1985, 102 pp. In the last 10 years, the domestication of jojoba, a little-known North American desert shrub, has been all but completed. This report describes the plant and its promise to provide a unique vegetable oil and many likely industrial uses. ISBN 0-309-04251-8.

63. **Quality-Protein Maize.** 1988, 130 pp. Identifies the promise of a nutritious new form of the planet's third largest food crop. Includes chapters on the importance of maize, malnutrition and protein quality, experiences with quality-protein maize (QPM), QPM's potential uses in feed and food, nutritional qualities, genetics, research needs, and limitations. ISBN 0-309-04262-3.

64. **Triticale: A Promising Addition to the World's Cereal Grains.** 1988, 105 pp. Outlines the recent transformation of triticale, a hybrid between wheat and rye, into a food crop with much potential for many marginal lands. The report discusses triticale's history, nutritional quality, breeding, agronomy, food and feed uses, research needs, and limitations. ISBN 0-309-04263-1.

67. **Lost Crops of the Incas.** 1989, 415 pp. The Andes is one of the seven major centers of plant domestication but the world is largely unfamiliar with its native food crops. When the Conquistadores brought the potato to Europe, they ignored the other domesticated Andean crops—fruits, legumes, tubers, and grains that had been cultivated for centuries by the Incas. This book focuses on 30 of the "forgotten" Incan crops that show promise not only for the Andes but for warm-temperate, subtropical, and upland tropical regions in many parts of the world. ISBN 0-309-04264-X.

70. **Saline Agriculture: Salt-Tolerant Plants for Developing Countries.** 1989, 150 pp. The purpose of this report is to create greater awareness of salt-tolerant plants and the special needs they may fill in developing countries. Examples of the production of food, fodder, fuel, and other products are included. Salt-tolerant plants can use land and water unsuitable for conventional crops and can harness saline resources that are generally neglected or considered as impediments to, rather than opportunities for, development. ISBN 0-309-04266-6.

74. **Vetiver Grass: A Thin Green Line Against Erosion.** 1993, 182 pp. Vetiver is a little-known grass that seems to offer a practical solution for controlling soil loss. Hedges of this deeply rooted species catch and hold back sediments. The stiff foliage acts as a filter that also slows runoff and keeps moisture on site, allowing crops to thrive when neighboring ones are desiccated. In numerous tropical locations, vetiver hedges have restrained erodible soils for decades and the grass—which is pantropical—has shown little evidence of weediness. ISBN 0-309-04269-0.

Innovations in Tropical Forestry

35. **Sowing Forests from the Air.** 1981, 64 pp. Describes experiences with establishing forests by sowing tree seed from aircraft. Suggests testing and development of the techniques for possible use where forest destruction now outpaces reforestation. ISBN 0-309-04257-7.

41. **Mangium and Other Fast-Growing Acacias for the Humid Tropics.** 1983, 63 pp. Highlights 10 acacia species that are native to the tropical rain forest of Australasia. That they could become valuable forestry resources elsewhere is suggested by the exceptional performance of *Acacia mangium* in Malaysia. ISBN 0-309-04165-1.

42. **Calliandra: A Versatile Small Tree for the Humid Tropics.** 1983, 56 pp. This Latin American shrub is being widely planted by villagers and government agencies in Indonesia to provide firewood, prevent erosion, provide honey, and feed livestock. ISBN 0-309-04166-X.

43. **Casuarinas: Nitrogen-Fixing Trees for Adverse Sites.** 1983, 118 pp. These robust, nitrogen-fixing, Australasian trees could become valuable resources for planting on harsh eroding land to provide fuel and other products. Eighteen species for tropical lowlands and highlands, temperate zones, and semiarid regions are highlighted. ISBN 0-309-04167-8.

52. **Leucaena: Promising Forage and Tree Crop for the Tropics.** 1984 (2nd edition), 100 pp. Describes a multipurpose tree crop of potential value for much of the humid lowland tropics. Leucaena is one of the fastest growing and most useful trees for the tropics. ISBN 0-309-04250-X.

71. **Neem: A Tree for Solving Global Problems.** 1992, 151 pp. The neem tree is potentially one of the most valuable of all trees. It shows promise for pest control, reforestation, and improving human health. Safe and effective pesticides can be produced from seeds. Neem can grow in arid and humid tropics and is a fast-growing source of fuelwood. ISBN 0-309-04686-6.

Managing Tropical Animal Resources

32. **The Water Buffalo: New Prospects for an Underutilized Animal.** 1981, 188 pp. The water buffalo is performing notably well in recent trials in such unexpected places as the United States, Australia, and Brazil. Report discusses the animal's promise, particularly emphasizing its potential for use outside Asia. ISBN 0-309-04159-7.

44. **Butterfly Farming in Papua New Guinea.** 1983, 36 pp. Indigenous butterflies are being reared in Papua New Guinea villages in a formal government program that both provides a cash income in remote rural areas and contributes to the conservation of wildlife and tropical forests. ISBN 0-309-04168-6.

45. **Crocodiles as a Resource for the Tropics.** 1983, 60 pp. In most parts of the tropics, crocodilian populations are being decimated, but programs in Papua New Guinea and a few other countries demonstrate that, with care, the animals can be raised for profit while protecting the wild populations. ISBN 0-309-04169-4.

46. **Little-Known Asian Animals with a Promising Economic Future.** 1983, 133 pp. Describes banteng, madura, mithan, yak, kouprey, babirusa, javan warty pig, and other obscure but possibly globally useful wild and domesticated animals that are indigenous to Asia. ISBN 0-309-04170-8.

68. **Microlivestock: Little-Known Small Animals with a Promising Economic Future.** 1990, 449 pp. Discusses the promise of small breeds and species of livestock for Third World villages. Identifies more than 40 species, including miniature breeds of cattle, sheep, goats, and pigs; eight types of poultry; rabbits; guinea pigs and other rodents; dwarf deer and antelope; iguanas; and bees. ISBN 0-309-04265-8.

Health

49. **Opportunities for the Control of Dracunculiasis.** 1983, 65 pp. Dracunculiasis is a parasitic disease that temporarily disables many people in remote, rural areas in Africa, India, and the Middle East. Contains the findings and recommendations of distinguished scientists who were brought together to discuss dracunculiasis as an international health problem. ISBN 0-309-04172-4.

55. **Manpower Needs and Career Opportunities in the Field Aspects of Vector Biology.** 1983, 53 pp. Recommends ways to develop and train the manpower necessary to ensure that experts will be available in the future to understand the complex ecological relationships of vectors with human hosts and pathogens that cause such diseases as malaria, dengue fever, filariasis, and schistosomiasis. ISBN 0-309-04252-6.

60. **U.S. Capacity to Address Tropical Infectious Diseases.** 1987, 225 pp. Addresses U.S. manpower and institutional capabilities in both the

public and private sectors to address tropical infectious disease problems. ISBN 0-309-04259-3.

Resource Management

50. **Environmental Change in the West African Sahel.** 1984, 96 pp. Identifies measures to help restore critical ecological processes and thereby increase sustainable production in dryland farming, irrigated agriculture, forestry and fuelwood, and animal husbandry. Provides baseline information for the formulation of environmentally sound projects. ISBN 0-309-04173-2.

51. **Agroforestry in the West African Sahel.** 1984, 86 pp. Provides development planners with information regarding traditional agroforestry systems—their relevance to the modern Sahel, their design, social and institutional considerations, problems encountered in the practice of agroforestry, and criteria for the selection of appropriate plant species to be used. ISBN 0-309-04174-0.

72. **Conserving Biodiversity: A Research Agenda for Development Agencies.** 1992, 127 pp. Reviews the threat of loss of biodiversity and its context within the development process and suggests an agenda for development agencies. ISBN 0-309-04683-1.

Forthcoming Books from BOSTID

Lost Crops of Africa: Volume 1, Grains. (1992) Many people predict that Africa in the near future will not be able to feed its projected population. What is being overlooked, however, is that Africa has more than 2,000 native food plants, almost none of which are receiving research or recognition. This book describes the potentials of finger millet, fonio, pearl millet, sorghum, tef, and other cereal grains with underexploited potential for Africa and the world.

Lost Crops of Africa: Volume 2, Cultivated Fruits. (1992) This report will detail the underexploited promise of about a dozen fruits that are native to Africa. Included will be such species as baobab, butter fruit, horned melon, marula, and watermelon.

Lost Crops of Africa: Volume 3, Wild Fruits. (1992) This study will highlight African native fruits that are promising candidates for domestication as well as for greater exploration in the wild. Included will be aizen, chocolate berries, gingerbread plums, monkey orange, raisin trees and more than 20 other species.

BOSTID Publication Distributors

U.S.:

AGRIBOOKSTORE
1611 N. Kent Street
Arlington, VA 22209

AGACCESS
PO Box 2008
Davis, CA 95617

Europe:

I.T. PUBLICATIONS
103–105 Southhampton Row
London WC1B 4HH
Great Britain

S. Toeche-Mittler
TRIOPS Department
Hindenburgstr. 33
6100 Darmstadt
Germany

T.O.O.L. PUBLICATIONS
Sarphatistraat 650
1018 AV Amsterdam
Netherlands

Asia:

ASIAN INSTITUTE OF TECHNOLOGY
Library & Regional
Documentation Center
PO Box 2754
Bangkok 10501
Thailand

NATIONAL BOOKSTORE
Sales Manager
PO Box 1934
Manila
Philippines

UNIVERSITY OF MALAYA COOP. BOOKSHOP LTD.
Universiti of Malaya
Main Library Building
59200 Kuala Lumpur
Malaysia

RESEARCHCO PERIODICALS
1865 Street No. 139
Tri Nagar
Delhi 110 035
India

CHINA NATL PUBLICATIONS
IMPORT & EXPORT CORP.
PO Box 88F
Beijing
China

South America:

ENLACE LTDA.
Carrera 6a. No. 51-21
Bogota, D.E.
Colombia

Africa:

TAECON
c/o Agricultural Engineering Dept
P.O. Box 170 U S T
Kumasi
Ghana

Australasia:

TREE CROPS CENTRE
P.O. Box 27
Subiaco, WA 6008
Australia

ORDER FORM

Please indicate on the labels below the names of colleagues, institutions, libraries, and others that might be interested in receiving a copy of Vetiver: A Thin Green Line Against Erosion.

Please return this form to:

Vetiver Report, FO 2060V
National Academy of Sciences
2101 Constitution Avenue N.W.
Washington, D.C. 20418, USA